# Spacecraft Charging

# Spacecraft Charging

**EDITED BY**

**Shu T. Lai**
*Senior Research Physicist*
*Space Vehicles Directorate*
*Air Force Research Laboratory*
*Hanscom Air Force Base, Massachusetts*

**Volume 237**
**Progress in Astronautics and Aeronautics**

**Frank K. Lu, Editor-in-Chief**
*University of Texas at Arlington*
*Arlington, Texas*

Published by
American Institute of Aeronautics and Astronautics, Inc.
1801 Alexander Bell Drive, Reston, VA 20191-4344

American Institute of Aeronautics and Astronautics, Inc., Reston, Virginia

1  2  3  4  5

# PROGRESS IN ASTRONAUTICS AND AERONAUTICS

## EDITOR-IN-CHIEF

### Frank K. Lu
*University of Texas at Arlington*

Because our society is so heavily dependent on satellites for many of the conveniences it expects such as GPS, mobile communications, weather maps, and so forth, there is an expectation that these systems should function smoothly. The challenge to the spacecraft community is to ensure that these expectations are met. Thus, surviving the electromagnetic environment of space intact, with prolonged years of service, is of great concern in the design of spacecraft, perhaps more so than of terrestrial systems. It is with this backdrop that I welcome this newest volume to the Progress Series. Dr. Shu Lai, through the auspices of the AIAA Atmospheric and Space Environment Technical Committee, has gathered top experts to contribute their insights into the latest ideas on dealing with spacecraft charging. The contributions provide an overview of spacecraft charging, as well as fundamental physics in an easy-to-understand form that will be useful to all practitioners as well as newcomers in this field.

# TABLE OF CONTENTS

Shu T. Lai, *U.S. Air Force Research Laboratory, Hanscom Air Force Base, Massachusetts*

Shu T. Lai, *U.S. Air Force Research Laboratory, Hanscom Air Force Base, Massachusetts*

Dale C. Ferguson, *U.S. Air Force Research Laboratory, Kirtland Air Force Base, New Mexico* and G. Barry Hillard, *NASA Glenn Research Center, Brookpark, Ohio*

## Chapter 4   Surface Discharge on Spacecraft   . . . . . . . . . . . . . . .  **75**

Mengu Cho, *Kyushu Institute of Technology, Kitakyushu, Japan*

## Chapter 5   Spacecraft Charging Simulation   . . . . . . . . . . . . . . . . .  **99**

Mengu Cho, *Kyushu Institute of Technology, Kitakyushu, Japan*

## Chapter 6   Spacecraft Charging in the Auroral Oval   . . . . . . . . .  **125**

Lars Eliasson, *Swedish Institute of Space Physics, Kiruna, Sweden* and Anders I. Eriksson, *Swedish Institute of Space Physics, Uppsala, Sweden*

## Chapter 7   Internal Charging .......................... 143

D. J. Rodgers and J. Sørensen, *ESA/ESTEC, Noordwijk, The Netherlands*

# PREFACE

Spacecraft charging affects scientific measurements onboard spacecraft and poses hazards to delicate electronic instruments. It may affect the telemetry and navigation systems. In severe cases, it may be destructive to the spacecraft. It is often necessary to monitor the level of spacecraft charging on spacecraft for measurement and safety purposes.

In recent years, it has been recognized that the sun controls the near-Earth space weather. Under the influence of solar ejections, the space plasmas can become hot and the space weather stormy. Indeed, space plasmas are dynamic. The prediction of the varying space weather in near-Earth space is now a hot research topic at NASA and research universities worldwide.

Once charging occurs, the ambient ions and electrons will take part in the current balance, governing the exact spacecraft potential. When different parts of a spacecraft charge to different potentials, potential barriers and wells can form near the surfaces. In low Earth orbits, the spacecraft velocity exceeds the average ambient ion velocity, thus forming wakes behind spacecraft. With emissions of artificial electron beams or ion beams, a spacecraft can charge to high potentials. Spacecraft charging in the Van Allen radiations belts involves high-energy (million electron volts) electrons and protons penetrating into dielectric materials. Cumulative fluence of high-energy electrons and protons can cause high internal electric fields, leading to sudden breakdowns. There are well-tested mitigation techniques for spacecraft charging, but each one has its advantages and disadvantages.

After some twenty years of development, spacecraft charging is now becoming a field of its own. Its importance in relation to the dynamic space weather, effects on scientific measurements, and spacecraft anomalies is now recognized. Many spacecraft will be launched in this millennium, and some will travel to the moon, Mars, and other planets. The harsh environments will pose spacecraft charging problems for spacecraft in the future. This field is gaining momentum internationally as witnessed by the proliferation of sponsorships and organizations of spacecraft charging technology conferences in recent years. More and more universities are teaching spacecraft charging. There is timely need for a state-of-the-art book on the progress in various key aspects of spacecraft charging.

A few years ago, the AIAA Atmospheric and Space Environment Technical Committee (ASETC) realized the need and importance for such a book. The AIAA Atmospheric and Space Committee on Standards (ASECOS) approved the book project with the book title, *A Guide to Spacecraft Charging and Mitigation*. As the book project evolved, two chapters were written by the volume editor while the other five chapters were written by some of the top experts on various key aspects of spacecraft charging. Eventually, AIAA decided that this book should be titled as *Spacecraft Charging*.

This book is written assuming that the readers have no or little background in the field of spacecraft charging in space plasma environments and particle interaction physics. It is our primary intention that *Spacecraft Charging* would serve as a guide and a reference in the field.

**Shu T. Lai**
*Hanscom Air Force Base, Massachusetts*
*August 2011*

# ACRONYMS AND ABBREVIATIONS

| | |
|---|---|
| AFRL | U.S Air Force Research Laboratory |
| APSA | advance photovoltaic solar array |
| BEY | backscattered electron yield |
| CFRP | carbon-fiber-reinforced plastic |
| CHAWS | charging hazards and wake studies |
| CME | coronal mass ejection |
| CRRES | Combined Release and Radiation Effects Satellite |
| DDC | deep dielectric charging |
| DICTAT | Dielectric Internal Charging Threat Assessment Tool |
| DMSP | Defense Meteorological Satellite Program |
| EAPU | electric auxiliary power unit |
| ELF | electron emitting film |
| EM | electromagnetic |
| EMI | electromagnetic interference |
| EMU | extravehicular maneuvering unit (spacesuit) |
| EOIM | effects of oxygen interaction with materials |
| EOS-AM1 | Earth-observing system—Morning Side number 1 (renamed Terra after launch) |
| EPSAT | Environmental Power System Analysis Tool |
| ESADDC | European Space Agency Deep Dielectric Charging |
| ESD | electrostatic discharge |
| ESTEC | European Space Research and Technology Centre |
| Eureca | European retrievable carrier |
| EUV | extreme ultraviolet |
| EVA | extravehicular activity |
| EWB | Environments WorkBench |
| FEF | field enhancement factor |
| FPP | floating potential probe |
| GEANT4 | geometry and tracking |
| GEO | geosynchronous Earth orbit |
| GES | geospace environment simulation |
| GRC | Glenn Research Center |
| GTO | geostationary transfer orbit |
| GUI | graphical user interface |
| IDM | internal discharge monitor |
| IPG | inverted potential gradient |
| IRI | International Reference Ionosphere |
| ISS | International Space Station |
| ITAR | International Traffic-in-Arms Regulations |
| ITO | indium tin oxide |
| ITS | integrated Tiger Series |

LANL            Los Alamos National Laboratory
LDEF            long-duration exposure facility
LEO             low Earth orbit
LHC             Large Hadron Collider
MEMS            microelectromechanical systems
MEO             middle Earth orbit
MET             Marshall Engineering Thermosphere
MIRIAD          Module Integrator and Rule-Based Intelligent Analysis Database
MLT             magnetic local time
MSFC            Marshall Spaceflight Center
MSIS            mass spectrometer incoherent scatter
NASCAP          NASA Charging Analyzer Program
NASCAP-2K       NASA/Air Force Spacecraft Charging Analyzer Program
NOAA            National Oceanographic and Atmospheric Administration
NPG             normal potential gradient
NSA             nonsustained arc
OML             orbital motion limited
PAS-6           Space Systems/Loral Commercial communications satellite
PASP Plus       photovoltaic array space power plus diagnostics
PCU             plasma contactor unit
PEO             polar Earth orbit
PIC             particle in cell
PIM             plasma interactions model
PIX             Plasma Interactions Experiment
PIX-II          Plasma Interactions Experiment – II
PMAD            power management and distribution
PMG             plasma motor generator
ProSEDS         Propulsive Small Expendable Deployer System
PSA             permanent sustained arc
PT              particle tracking
RC              resistance-capacitance
RCS             reaction control system (attitude thrusters)
RIC             radiation-induced conductivity
RTV             room-temperature vulcanized rubber
SAMPIE          Solar-Array Module Plasma Interactions Experiment
SCATHA          spacecraft charging at high altitudes
SEC             surface emission cathode
SEY             secondary electron yield
SMP             symmetric multiple processor
SPEAR           space power experiments aboard rockets
SPENVIS         European Space Agency Space Environment Information System
SPIS            Spacecraft Plasma Interaction System
SREM            standard radiation environment monitor
SS              Space Systems

| | |
|---|---|
| SSULI | Special Sensor Ultraviolet Limb Imager |
| Tempo-2 | Space Systems/Loral Commercial Communications Satellite |
| TSA | temporary sustained arc |
| TSS | tethered satellite system |
| TSS-1R | tethered satellite system—first reflight |
| USAF | U.S. Air Force |
| UV | ultraviolet |

# NOMENCLATURE

| Item | Definition |
|------|-----------|
| $A$ | surface area, m$^2$; coefficient in the backscattered electron formula |
| $A_{body}$ | total surface area of satellite body, m$^2$ |
| $A_{panel}$ | total area of solar panel, m$^2$ |
| $A_{patch}$ | area of insulator patch, m$^2$ |
| $a$ | straggle distance |
| $B$ | coefficient in the backscattered electron formula; magnetic field, T |
| $C$ | capacitance; coefficient in the backscattered electron formula |
| $C_{cg}$ | capacitance per unit area of coverglass with respect to spacecraft body, F/m$^2$ |
| $C_d$ | capacitance per unit area of insulator surface with respect to spacecraft body, F/m$^2$ |
| $C_{sat}$ | capacitance between satellite body and plasma, F |
| $D$ | radiation dose rate, Gy/s |
| $d$ | thickness; insulator thickness, m |
| $E$ | primary electron energy; electric field, V/m |
| $E_e$ | electron incident energy, eV |
| $E_{max}$ | primary electron energy at which the secondary electron yield is maximum |
| $E_0$ | parameter specifying the enhancement fall-off rate of $\eta$, which is material specific |
| $E_2$ | second crossing energy of $\delta(E)$ |
| $e$ | charge of the electron; $1.6 \times 10^{-19}$, C |
| $f(E)$ | electron velocity distribution expressed in terms of electron energy |
| $I$ | current, A |
| $I_a$ | active emission current, A |
| $I_{be}$ | backscattered electron current, A |
| $I_c$ | leak current from an insulator surface, A |
| $I_e$ | ambient electron current, A |
| $I_e(\phi)$ | electron current as a function of surface potential $\phi$ |
| $I_i$ | ambient ion current, A |
| $I_i(\phi)$ | ion current as a function of surface potential $\phi$ |
| $I_{ph}$ | photoelectron current, A |
| $I_{se}$ | electron-induced secondary electron current, A |
| $I_{si}$ | ion-induced secondary electron current, A |
| $J$ | flux |
| $J_k$ | flux with subscript $k$ labeling the type of flux |
| $J_0$ | electron thermal current density, A/cm$^2$ |
| $j$ | current density |
| $j_{be}$ | backscattered electron current density, A/m$^2$ |
| $j_c$ | leakage current density through the insulator, A/m$^2$ |

| | |
|---|---|
| $j_e$ | ambient electron current density, A/m$^2$ |
| $j_{es}$ | electron-induced electron current density, A/m$^2$ |
| $j_{oe}$ | electron current density in the far field, A/m$^2$ |
| $j_{oi}$ | ion current density in the far field, A/m$^2$ |
| $j_{ph}$ | photoelectron current density, A/m$^2$ |
| $j_s$ | leakage current density through the insulator, A/m$^2$ |
| $j_{si}$ | ion-induced electron current density, A/m$^2$ |
| $k$ | Boltzmann's constant, $= 1.38 \times 10^{-23}$, J/K |
| $l$ | length, m |
| $m$ | electron mass |
| $m_e$ | electron mass, $= 9.1 \times 10^{-31}$, kg |
| $m_i$ | ion mass, kg |
| $n$ | electron density |
| $n, N$ | number density, cm$^{-3}$ or m$^{-3}$ |
| $n_e$ | electron density, m$^{-3}$ |
| $n_i$ | ion density, m$^{-3}$ |
| $P$ | power, W |
| $Q$ | electric charge |
| $q$ | charge, C |
| $q_e$ | electron charge |
| $q_i$ | ion charge |
| $R$ | sphere radius, m; range; resistance |
| $R_s$ | surface resistivity, $\Omega$/sq |
| $S$ | stopping power |
| $s$ | parameter of surface condition |
| $T$ | electron temperature |
| $T_e$ | electron temperature ($kT_e$ in eV or K) |
| $T_i$ | ion temperature ($kT_i$ in eV or K) |
| $T^*$ | critical electron temperature for the onset of spacecraft charging |
| $V$ | voltage or potential, V |
| $V_a$ | solar-array generating voltage, V |
| $V^*$ | critical voltage |
| $v$ | electron velocity; velocity, m/s |
| $\alpha$ | exponent in the Mott–Smith Langmuir attraction term |
| $\beta$ | energy |
| $\beta^*$ | critical energy |
| $\gamma_{ee}$ | secondary electron emission yield |
| $\Delta V$ | differential voltage, V |
| $\Delta \eta$ | additional term for modifying the BSY formula of [2] from Chap. 2 |
| $\delta$ | secondary electron yield (also called secondary electron emission coefficient) |
| $\delta_F$ | $\delta$ formula of Furman [12] from Chap. 2 |
| $\delta_K$ | $\delta$ formula of Katz et al. [9] from Chap. 2 |
| $\delta_L$ | $\delta$ formula of Lin and Joy [11] from Chap. 2 |

| | |
|---|---|
| $\delta_m$ | maximum value of $\delta(E)$ |
| $\delta_{max}$ | maximum value of secondary electron yield |
| $\delta_S$ | $\delta$ formula of Sanders and Inouye [7] from Chap. 2 |
| $\delta_{SZ}$ | $\delta$ formula of Scholtz [10] from Chap. 2 |
| $\varepsilon$ | permittivity |
| $\varepsilon_o$ | dielectric constant of vacuum, $= 8.85 \times 10^{-12}$, F/m |
| $\eta$ | backscattered electron yield |
| $\eta_K$ | backscattered electron yield of Katz et al. [9] from Chap. 2 |
| $\theta$ | electron incident angle, rad |
| $\lambda_D$ | Debye length, m or cm |
| $\rho$ | charge density, $C/m^2$ |
| $\rho_b$ | bulk material resistivity of insulator, $\Omega m$ |
| $\sigma$ | electrical conductivity, $1/\Omega/m$ |
| $\sigma_o$ | dark conductivity, $1/\Omega/m$ |
| $\tau$ | time constant |
| $\phi$ | spacecraft surface potential, V |
| $\phi_{cg}$ | coverglass potential, V |
| $\phi_d$ | insulator potential, V |
| $\phi_s$ | sphere potential, V |
| $\phi_{sc}$ | spacecraft body potential, V |
| $\varphi$ | potential, V |

# CHAPTER 1

# Overview of Surface and Deep Dielectric Charging on Spacecraft

## Shu T. Lai*

*U.S. Air Force Research Laboratory, Hanscom Air Force Base, Massachusetts*

## 1.1 INTRODUCTION

Spacecraft charging electrostatic discharge (ESD) has caused by far the most environmentally related anomalies on spacecraft. Surface charging and deep dielectric charging have caused the most serious anomalies, that is, those that have resulted in the loss of mission [1] (see Fig. 1.1). ESD encompasses both surface and internal charging. The effects of ESD are a combination of the electron environment and its interaction with specific spacecraft surfaces and components.

## 1.2 OVERVIEW

There is no single technology that is particularly susceptible to ESD. ESD has two types of effects, that is, current flow through wires to sensitive instruments and electromagnetic (EM) wave interference to telemetry. Therefore, sensitive instruments and telemetry are susceptible to ESD. There is no single technology that mitigates all ESD conditions. A variety of mitigation techniques [2] exist.

Although the basic principles for most ESD mechanisms have been known for nearly 20 years, much more research is needed to better understand spacecraft charging and ESD. Surface and payload design should take expected levels of electron accumulation into account and mitigate them in the context of particular design susceptibilities and mission exposure.

Surface charging can generally be mitigated with careful attention to grounding design or by means of emission of ions, plasma, or polar molecules [3]. Deep dielectric charging can be mitigated to some extent by using materials of finite conductivity. Filters can be used for minimizing electromagnetic wave interference. Even with adequate design techniques, ESD can cause problems if the triggering events are larger than expected, or if surface damage is sustained from either the launch environment or on-orbit debris or meteorites.

---

*Senior Research Physicist, Space Vehicles Directorate, 01731. Associate Fellow AIAA.

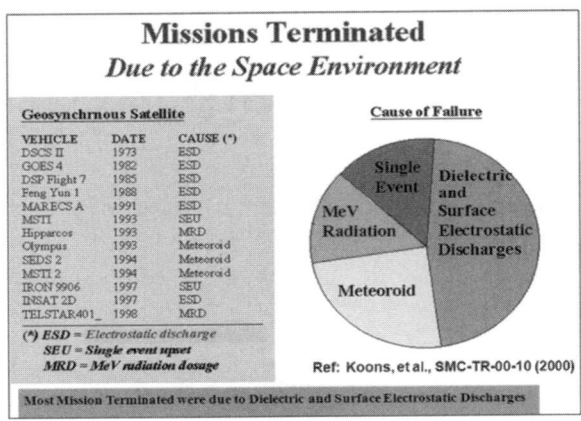

**FIG. 1.1    Missions terminated due to the space environment (data taken from [1]). Most missions terminated are due to surface and deep dielectric ESD.**

The capability of forecasting environmental conditions that can lead to an ESD event is a separate consideration. Internal charging in dielectrics often occurs one to several days after a major magnetic storm. Hence, the storm itself is a warning that a high level of energetic (MeV) electrons might be present in the radiation belts. Efforts to do these predictions have been undertaken using linear prediction filters and neural networks. These methods for predicting storms are inadequate because they assume a stationary time series.

For predicting significant magnetic storms, it is necessary to observe and predict the weather on the sun, which controls the space weather of the Earth. If a twisted magnetic field configuration, sigmoid signature, shows up on the sun, it is possible that a coronal mass ejection (CME) might erupt in a few hours [4]. A CME arrival at the Earth's magnetosphere can cause rapid increase of energetic (MeV) electrons, which in turn can cause internal charging.

Hardening of the ambient electron energy spectrum causes surface charging. A surge in the incoming electron flux [5] of about 20 to 40 keV and a rise of the ambient electron temperature [6–11] above a critical value of about a few kiloelectron volts, depending on the surface material, are manifestations of electron spectrum hardening signaling the onset of spacecraft surface charging. More on this topic will be discussed in Sec. 1.4.3.

Differential charging typically occurs over seconds to minutes depending on the capacitances involved. Capacitance $C$ depends on the surface area $A$, the permittivity $\varepsilon$, and the thickness $d$ of the material between surfaces ($C = \varepsilon A/d$). For a given surface size and material, it is the thickness that controls the capacitance. Most of the undesired effects of ESD include physical damage to sensitive electronics and electromagnetic wave or pulse generation. The magnitude of an ESD depends on the energy $\beta$ stored in a capacitance $C$:

$$\beta = \frac{CV^2}{2} = \frac{\varepsilon A V^2}{(2d)} \tag{1.1}$$

The critical energy $\beta^*$ above which damage can occur depends on the specifics of the system involved. Lacking any specifics, a value of $\beta^*$ at 10 mJ is advised. For each value of $\beta^*$, one can plot the critical voltage $V^*$ as a function of the thickness $d$. Very thin (small $d$) dielectric layers sandwiched between conducting surfaces are often sources of high-energy ESD. Thin thermal blankets have been repeatedly reported as sources of ESD-induced anomalies on satellites [1].

Leaky dielectrics, proper grounding, and shielding can be used to reduce the possibility of internal charging. Internal discharge occurs when the critical electric field $E^*$ reaches $10^6$ to $10^8$ V/m depending on the material [12]. To mitigate internal charge buildup, the best way known to date is to use materials that are partially conducting. One needs a nonconducting material for insulation purposes, but, on the other hand, one needs to have finite conductivity to avoid buildup of internal charge. The tradeoff depends on the requirement of the system. Another method for mitigating internal charging is to use shielding. Ions and electrons of different energies penetrate into materials to different depths [12]. Shielding can prevent electrons or ions below a certain desired energy from reaching an instrument. There is a tradeoff: too much shielding can blind the instruments.

There is yet another method, that is, using thin materials or instruments so that high-energy electrons or ions pass completely through, without leaving behind much deposit inside. It is difficult, but possible, to build instruments that are very thin.

## 1.3  HISTORICAL DEVELOPMENT

Tree rings indicate the history of tree growth over the years. The Spacecraft Charging Technology Conferences can be regarded as a kind of tree ring reflecting the growth of this field.

The field of spacecraft charging was practically nonexistent before the mid-1970s. It began to take shape with the First Spacecraft Charging Technology Conference, sponsored by U.S Air Force Research Laboratory (AFRL) and NASA and held at the U.S. Air Force Academy, Colorado Springs, Colorado, in 1978. AFRL and NASA have been sponsors of each of the Spacecraft Charging Technology Conference ever since.

The first four conferences were all held at the U.S. Air Force Academy in Colorado. The fifth one was at the Naval Postgraduate School, Monterey, California. The sixth was at AFRL, Hanscom Air Force Base, Massachusetts. The seventh conference began a tradition of rotating international sponsorship. The European Space Agency Technology Centre's (ESTEC) sponsorship of the Seventh Conference, held in Noordwijk, the Netherlands, set a significant milestone in the history of the conference series, making it a truly international event. The eighth conference was at the NASA Marshall Space Flight Center, Huntsville, Alabama. The ninth conference was held at Tsukuba, Japan, in 2005. Each one of the conferences in the series has been a great success. The topical discussions have greatly helped in making progress in the fields of spacecraft charging, spacecraft interactions, and related research areas. The tenth conference in this series was held in Biarritz, France, in 2007.

For the first time, the papers presented at a Spacecraft Charging Technology Conference were submitted to *IEEE Transactions on Plasma Science* and were published, after a rigorous referee process, in a special issue in Oct. 2006. The second in this series of Spacecraft Charging Technology Conference publications was published in the same journal, Oct. 2008.

## 1.4   SOME SPACECRAFT CHARGING TOPICS AND TRENDS

We will discuss some spacecraft charging topics and research trends in various areas in what follows. Critical comments in some areas are offered.

### 1.4.1   SPACECRAFT AS A LANGMUIR PROBE

In the early years of spacecraft charging research, much attention was on the uniform charging of a spherical spacecraft. It was realized that a spacecraft in a space plasma behaves like a Langmuir probe. In the laboratory, the voltage applied to a Langmuir probe causes the currents collected to vary in response. In space, the ambient currents cause the spacecraft voltage to vary in response. Laframboise and Parker [13] did analytical and numerical work on current collection by a Langmuir probe; their work was applicable to spacecraft charging studies. Whipple [14] studied potential barriers by using the ATS-5 and ATS-6 satellites. There was not much other satellite charging data available until 1979.

### 1.4.2   SCATHA SATELLITE

In 1979, AFRL and NASA launched the Spacecraft Charging at High Altitudes (SCATHA) Satellite, which was dedicated to spacecraft charging research at geosynchronous altitudes. SCATHA carried instruments for several different experiments on board. The launch was monumental, for it ushered in the dawn of spacecraft charging studies.

The SCATHA results agreed fairly with the general concept and understanding of spacecraft charging at that time. The charging level of SCATHA at geosynchronous altitudes was higher during geomagnetic storms than during quiet time. For the first time, voltage measurements of four different materials, isolated on the same spacecraft, confirmed the existence of differential charging [15].

Olsen et al. [16] studied the SCATHA data and found evidence of potential barriers and potential wells on the differentially charged spacecraft. They also studied effects of electron and ion-beam emissions. SCATHA was equipped with electron and ion guns (or beam emitters). Emitting the beams at various known beam energies and currents during geomagnetic conditions characterized by $kp$ (geomagnetic activity index), which was also known, provided excellent case studies of spacecraft charging. These case studies revealed much underlying physics governing spacecraft charging during charged particle beam emissions [3, 17–20].

Spacecraft surface charging is caused by the surface interaction with the plasma up to a few tens of kiloelectron volts in energy [21]. The broad advances by the many researchers using the SCATHA data have established spacecraft charging as a field of its own. Some of the salient features in the SCATHA research results have been surveyed by Mizera [22].

### 1.4.3 ONSET OF SPACECRAFT CHARGING AT GEOSYNCHRONOUS ALTITUDES

Hardening of the ambient electron energy spectrum causes spacecraft surfaces to charge to negative voltage. Hardening an electron spectrum means increasing the higher-energy electron population. Two typical phenomena have been observed at the onset of spacecraft charging. They are 1) rising ambient electron flux in the 20–40 keV range and 2) rising ambient electron temperature. Rising electron flux during a severe charging event observed on the SCATHA satellite has been documented [5]. Critical ambient electron temperature $T^*$ [6–10] has theoretical foundation, and new evidence of the existence of $T^*$ has been documented recently [11, 23] (see Table 1.1).

Current balance determines the surface potential. Because electrons are much faster than ions, the ambient electrons, not the ions, are the players determining the onset of spacecraft charging to negative voltages. Incoming ambient electrons generate outgoing (secondary and backscattered) electrons from the surface. For low-energy incoming ambient electrons, the number of outgoing electrons exceeds that of the incoming electrons, depending on the surface material. The high-energy ambient electrons are responsible for spacecraft charging. The two electron populations compete with each other. Therefore, when the ambient electron distribution hardens, spacecraft charging occurs.

In a Maxwellian plasma distribution, increasing the higher-energy electron population raises the electron temperature $T$. Obviously, there must exist a critical temperature $T^*$ above which the higher electron population wins the competition and spacecraft charging to negative voltage occurs.

Observations have shown that in geosynchronous orbit (GEO) conditions, the onset of spacecraft charging occurs at the critical temperature of the plasma electrons. Below it, there is no charging; above it, the charging level increases with the electron temperature. This handy parameter, critical temperature, can be useful for predicting surface charging in the geosynchronous environment, where most communication and surveillance satellites are and where the space plasma varies rapidly from a few electron volts to tens of kiloelectron volts in minutes or hours. Note that the numerical results of the critical temperature as well as

**TABLE 1.1   CRITICAL TEMPERATURE $T^*$ FOR THE ONSET OF SPACECRAFT CHARGING AT GEOSYNCHRONOUS ALTITUDES (DATA FROM [7])**

| Material | Critical Temperature $T^*$, keV | |
|---|---|---|
| | Isotropic | Normal |
| Magnesium | 0.4 | — |
| Aluminum | 0.6 | — |
| Kapton | 0.8 | 0.5 |
| Aluminum oxide | 2.0 | 1.2 |
| Teflon™ | 2.1 | 1.4 |
| Copper-beryllium | 2.1 | 1.4 |
| Glass | 2.2 | 1.4 |
| Silicon oxide | 2.6 | 1.7 |
| Silver | 2.7 | 1.2 |
| Magnesium oxide | 3.6 | 2.5 |
| Indium oxide | 3.6 | 2.0 |
| Gold | 4.9 | 2.9 |
| Copper-beryllium (activated) | 5.3 | 3.7 |
| Magnesium fluoride | 10.9 | 7.8 |

the spacecraft potential depend on the secondary and backscattered electron coefficients [24], which can be affected by the surface condition—smoothness, contamination, and surface temperature.

For ambient electron distributions deviating from a Maxwellian distribution, analytical solutions of the critical temperature formulation are generally difficult. Numerical calculations are possible. For the kappa distribution [25, 26], one can calculate the critical kappa temperature. Because the kappa temperature is related to the Maxwellian temperature, one can then calculate the corresponding critical temperature in the Maxwellian sense. Harris [27] has calculated the kappa temperature for a number of spacecraft charging onset cases using the Los Alamos National Laboratory (LANL) geosynchronous satellite data. He concluded that, at the onset, the critical Maxwellian temperature deduced from the critical kappa temperature is not significantly different from that obtained from the Maxwellian model directly. For high-level charging well beyond the onset, the kappa distribution can become significantly different from being Maxwellian. However, the concept of temperature is undefined for arbitrary distributions.

In the theoretical framework of Maxwellian plasmas, the two phenomena, 1) rising electron flux in a range of about 20–40 keV and 2) rising electron temperature beyond a critical temperature $T^*$, are consistent with each other [28]. However, method 1 is purely empirical, as it neglects surface material properties, whereas method 2 is "rigorous," within the Maxwellian approximation. The advantage of method 1 is simplicity, but the disadvantage is that sometimes it can produce "false positives," that is, the method always gives a surge in flux for large temperature $T$, but if $T$ is less than $T^*$, there is no charging. The two methods are not equivalent, unless $T$ exceeds $T^*$.

### 1.4.4    LANL GEOSYNCHRONOUS SATELLITES

At the turn of this century, the Los Alamos National Laboratory (LANL) geosynchronous satellite charging data became available. These were coordinated measurement data of the charging voltage, ambient electron temperature, local time, magnetic field, altitude, etc. They enabled new progress in spacecraft charging. For example, the data showed the existence of a critical temperature for the onset of spacecraft charging [11, 23].

There are shortcomings in the LANL data. The surfaces materials are not given, making accurate quantitative verification of spacecraft charging theories impossible. Also, the surface potential measurements data were obtained at one spot only. The location of the spot is not given. If there were many spots at known locations distributed all over the satellite, the data obtained would allow better verification of monopole-multipole distributions [29], bootstrap charging [30–33], satellite spin effects [34], sunlight and shadow effects [35], etc.

### 1.4.5    SPACECRAFT CHARGING IN THE IONOSPHERE

Spacecraft charging occurs not only at geosynchronous altitudes. It is most important there, as a consequence of the high-temperature plasmas during adverse space weather. However, spacecraft charging also occurs in the ionosphere, but generally at insignificantly lower voltages. The space shuttle has never experienced natural charging beyond a volt. Low-altitude spacecraft charging can occur in the wake of shields or with the operation of tethers. The Tethered Satellite System (TSS-1R) experiments and their results have been published [36]. Charging in the ionosphere can be induced by emission of artificial charged particle beams or by the presence of exposed high-voltage conductors. Wake charging has been confirmed by satellite experiments in the ionosphere [37]. TSS-1 and TSS-1R showed that a large vehicle could be charged in low Earth orbit (LEO) by **vxB.l** effects, and unfortunately TSS-1R was destroyed by arcing as a result of its charging [38]. More on ionospheric spacecraft charging will be given in Chapter 3.

### 1.4.6    AURORAL CHARGING OF SPACECRAFT

Charging at auroral altitudes can occur. The Swedish Freja satellite has reported observations confirming such events [39], and auroral charging on ISS has recently been confirmed. During inverted-V events, the satellite traversing the auroral field lines experienced surface potentials [40] surging to negative hundreds of volts or even kilovolts. These results suggest that charging in the auroral altitudes is hazardous and should not be taken lightly.

There are shortcomings of the Freja experiments [39]. For example, there was no measurement of the directions of the incoming ambient electrons and ions. The electrons and ions in that region are greatly affected by the converging magnetic field lines and the double-layer electric fields so that the incoming angles are nonisotropic. It would also be useful if there were instruments measuring the exact electron and ion distributions. With the knowledge of incoming angles and charge distributions, the researchers would be able to verify the existing charging theories with the observational results and perhaps even come up with new physical mechanisms to interpret the charging interactions. More on auroral charging will be given in Chapter 6.

### 1.4.7 COMPUTATIONS AND SIMULATIONS

The first comprehensive computational software for spacecraft charging research support was NASCAP [41–43] developed by Katz et al. [41], Mandell et al. [43], Stannard et al. [42], and their group, with the support by AFRL and NASA. The 1981 version had many limitations. Since 1981, there have been several variations, such as the NASCAP/Leo and the POLAR. The NASCAP-2K is a complete revision without the zero potential boundary assumption of the original NASCAP. It is probably the most comprehensive and best validated software for spacecraft charging computations at this time [44]. Unfortunately, NASCAP-2K is export controlled and cannot be shared with non-U.S. citizens.

The Environments WorkBench (EWB) tool [45] (data available online at http://see.msfc.nasa.gov/ee/model_ewb.htm) of NASA, a steady-state LEO modeling code for first-look LEO charging, has been used for modeling ISS [46] and other LEO spacecraft charging. An outgrowth of EWB, called PIM (plasma interactions model) is the official International Space Station (ISS) tool for charging by the high-voltage solar arrays (data available online at http://www.nasa.gov/mission_pages/station/science/experiments/Plasma_Interaction_Model.html).

The software SPIS (Spacecraft Plasma Interaction System) [47, 48] is being developed for calculating spacecraft surface charging in complex surface geometries. SPIS will compete with NASCAP-2K. User and developer documentations are available. The software is free. To access the software, the user needs to register for an account.

MUSCAT (Multi-Utility Spacecraft Charging Analysis Tool) [49] is a particle-in-cell code with particle tracking capabilities. The software uses parallel computation techniques for fast computation. It has good potential to rival NASCAP-2K, SPIS, and more. The software is available commercially.

It would be appropriate in the future to compare the effectiveness of the various computer models, or software codes, developed with different approaches.

More on computer code for spacecraft surface charging will be given in Chapter 5. Computer models for deep dielectric charging will be discussed in a later section.

## 1.4.8   LABORATORY MEASUREMENTS

In recent years, there have been advances in two areas of laboratory measurements in spacecraft charging. One area concerns measuring the secondary electron emission coefficient of various surface materials [50]. This is an important area because the secondary electron emission coefficient is central in determining the net electron current received by spacecraft surfaces. More comments on secondary and backscattered electrons will be given in Chapter 2.

Two critical comments on secondary electron yield are as follows. First, it is difficult to separate the secondary electrons from the backscattered electrons. Each of them has its own coefficients and its own energy range. Second, secondary electron yield depends on the surface condition. In another but related problem, photoelectron yield also depends on the surface condition. All outgoing electron currents, including secondary and backscattered electrons and photoelectrons, are very much affected by the local surface potentials. A small potential barrier of a few volts is sufficient to block many of the outgoing electrons [30–35].

Another area of laboratory measurements is in the study of charging, discharging, and avalanche arcing in coupons of solar cells. This area is very important because this is the only way to reveal the underlying physics and mechanisms in the solar-panel failures under space-like environments. Boeing, Utah State University, NASA Glenn Research Center (GRC) and Marshall Spaceflight Center (MSFC), and others in Japan and France are active in this area (see *IEEE Transactions in Plasma Science*, special issues Oct. 2006 and Oct. 2008).

A shortcoming of this type of research is that it has been too system oriented. For example, although it is significant to discover [51, 52] that the entire solar array discharges instead of part of the array only, the result might be system specific. In other words, if one changes the geometric arrangement of the solar cells, the results might be different. What one should seek is the basic physics, whenever one can. Papers by Vayner et al. [53, 54] are aimed at determining the basic physics involved.

Solar-panel arcing research activities are surging in Japan and France (for example, see [55–57]). The researchers in the United States, Japan, and France are very careful in distinguishing the characteristics of the initiation of arcs, the development of arcs, secondary arcs, sustaining arcs, and the avalanche ionization in arcing (for example, see [55–57]).

A shortcoming of this type of research is, again, too system oriented. Although it is useful to discover that there exists a critical voltage, say 500 V for initiating arcing in GEO, the critical voltage is a function of the distance of separation. What one should seek is the critical electric field and a list of coordinated parameters, enabling one to estimate whether Townsend's condition, which is more basic, for avalanche ionization is to occur. In cases where the separation is poorly defined, however, such as at the edges of solar cells, models must be used along with the system-dependent measurements to determine the critical electric fields. More on discharges will be given in the Chapter 4.

## 1.4.9  DEEP DIELECTRIC CHARGING

While the entire spacecraft charging community was paying attention to surface charging in the early 1980s, Robb Frederickson of AFRL was the lonely voice advocating deep dielectric charging (DDC). The natural fluxes of the ambient electrons and ions responsible for surface charging are orders of magnitude higher than those for deep dielectric charging. Surface charging responds almost instantaneously to the ambient flux temperature and can be easily measured. Deep dielectric charging is difficult to measure. The charges penetrated deep inside dielectrics accumulate over time, even though the surface voltage due to the deep charges can be very low.

The deep charges can stay inside insulators for many months (Frederickson, personal communication, 2000), and, as a result, build up strong electric fields. For cold dielectrics, the conductivity is much lower than for normal temperatures, and charges can build up for months or years. If the electric field exceeds a critical value depending on the material, dielectric breakdown (and therefore possible spacecraft anomalies) would occur. Violet and Frederickson [58] studied the spacecraft anomalies that occurred on the Combined Release and Radiation Effects Satellite (CRRES) in the radiation belts (i.e., the van Allen Belts), a deep dielectric charging space environment. Indeed, they found that anomalies occurred from time to time, but the surface potential measured during the anomalies was very low. The only physical explanation suggested [59] is that the electric fields of the deep charges are balanced by that of the charges of the opposite sign attracted to the shallow depth near the surface so that the total external electric field is nearly zero.

Another significant discovery [60] is the observation that an insulator irradiated by high-energy electrons in the laboratory emits plasmas and neutral gas. Obviously, the neutral gas emitted would help sustain arcs by electron-impact ionization, if arcing occurs.

Again, instead of obtaining results in system-specific cases only, one needs more systematic and basic studies in order to understand and interpret the physical mechanisms involved in various processes such as radiation-induced gas emissions, plasma emissions, and ionization chain reactions in sustained arcs.

In recent years, the Japan and French research teams have been successful in using acoustic methods for measuring in the laboratory the depth of the charge layer deposited by megaelectron-volt electron beams into dielectrics (for example, see [61]).

Deep dielectric charging should receive much more attention because it can cause internal damage to instruments located not only on spacecraft surfaces but also behind some spacecraft walls. Indeed, spacecraft anomalies in the radiation belts often occur during, or shortly after, high fluences of killer electrons [62–64] at high energies (MeVs). More on deep dielectric charging and spacecraft anomalies will be given in Chapter 7.

### 1.4.10  COMPUTATION AND SIMULATION FOR DDC

The ITS (Integrated Tiger Series) software uses Monte Carlo methods to calculate penetration of electrons, ions, and photons into materials (data available online at http://rsicc.ornl.gov/codes/ccc/ccc4/ccc-467.html). The software has very sophisticated capabilities and not only for calculating deep dielectric charging on spacecraft. ITS has been adopted in the software ESADDC (European Space Agency Deep Dielectric Charging) for calculating deposition and leakage of charges in dielectrics.

The DICTAT (Dielectric Internal Charging Threat Assessment Tool) [65, 66] software uses analytical equations for calculating currents and electric fields (data available online at http://www.space.qinetiq.com/idc/dictat2_sum_v0.0.pdf). It can assess the likelihood of dielectric breakdown on spacecraft. It has been adopted in the SPENVIS (European Space Agency Space Environment Information System) software of ESA for spacecraft deep dielectric charging calculations (data available online at http://www.spenvis.oma.be/spenvis/help/background/charging/dictat/dictatman.html).

The GEANT4 (Geometry and Tracking) software [67] uses Monte Carlo methods to simulate high-energy particle penetration into materials with many nuclear and high-energy particle interactions involved (data available online at http://geant4.web.cern.ch/geant4/). It is already useful in nuclear medicine and high-energy particle penetration areas. It is gaining popularity for simulating space radiation shielding and deep dielectric charging.

### 1.4.11  MITIGATION METHODS

There are mitigation methods for surface charging (Table 2). Active methods such as emitting electrons using electron guns, sharp spikes, or hot filaments are well known. Ironically, emitting a positive ion beam from the SCATHA satellite reduced the negative voltage level effectively [2]. Plasma emission [68] is the best active mitigation method known to date. When plasma is emitted from a high negatively charged spacecraft, the electrons escape while the positive ions return to the highly charged spots [3]. In recent years, ion emission has been used successfully for mitigation or for keeping the spacecraft potential at a desired level [69].

Passive methods do not require active controls. An example is using surface materials of high secondary emission coefficients. This method, relying on secondary electron emission, is of limited value because whenever the ambient electron distribution hardens, or the electron temperature exceeds a critical value depending on the surface material, charging would occur. Field emission from autonomous devices, such as sharp cones, can remove excess electrons from those spacecraft surfaces connected to the devices. In recent years, some notable advances in the development of field emission devices include the surface emission cathode (SEC) [70, 71] and the electron emitting film (ELF) [72]. An advantage of the field emission method is that it is passive and needs no command from the ground. Emitting electrons from a conducting surface connected to the device can reduce the negative voltage of the surface but cannot do so for surfaces electrically isolated from the emitting surface. Thus, the result might be differential charging. In differential charging, both the magnitude and the direction of the potential gradient are important. If the potential of a conducting surface is positive relative to that of its dielectric neighbor, the situation is called a normal gradient. Conversely, if the potential of the conducting surface is relatively negative, the situation is called an inverted gradient. A discharge is more probable in an inverted gradient because an electron current is easier to generate from a conducting surface than from a dielectric one.

Partially conductive paint such as indium oxide also helps. However, a uniform paint all over a spacecraft for avoiding differential charging would cover up the instruments and solar arrays, which are the eyes, ears, and mouth of the spacecraft. A critical overview of some spacecraft charging mitigation methods is given in [2].

Plasma contactors (Special issue, *Geophysical Research Letters*, Vol. 25, Nos. 4 – 5, 1998) are useful for mitigating voltage of both signs in the lower ionosphere. Positive voltage charging can occur when electrons are drawn from one part of a large spacecraft to another, from one end of a tether to the other, or from an artificial electron beam device. Negative charging on the ISS [46] is routinely mitigated during extravehicular activity (EVA) operations by the use of a plasma contactor. Plasma contactors have also been used on geosynchronous (GEO) spacecraft. A critical overview on some spacecraft charging mitigation techniques, including plasma contactor, for ISS is given [73].

For deep dielectric charging, there is no good mitigation method at this time. Three approaches have been used. One approach is to use shields. Indeed, shields can block high-energy electrons and ions from reaching sensitive instruments. However, covering up the eyes, nose, and ears is not a desirable way for conducting space observations and measurements. Another approach is to use very thin electronics, so that the high-energy electrons and ions would pass through without staying inside. The "thin" requirement hampers the design of sophisticated electronic instruments. A third approach is to use dielectric materials with adequate conductivity for reducing internal charge build up. Better mitigation methods are yet to be invented.

We should mention some mitigation methods that are common practice. They are nonspecific to any particular spacecraft charging mechanism. For example, install redundant circuits on spacecraft, ground all sensitive areas to the spacecraft chassis, avoid operations during severe geomagnetic storms and solar disturbance events, shield all cables properly, and do not expose wire connections, etc. Finally, if a single high-energy particle, perhaps from cosmic rays, hits the spacecraft and causes an upset event, it is difficult to prevent it. Fortunately, such hits are rare, and the disturbance might be local or temporary.

# REFERENCES

[1]  Koons, H. C., Mazur, J. E., Selesnick, R. S., Blake, J. B., Fennell, J. F., Roeder, J. L., and Anderson, P. C., "The Impact of Space Environment on Space Systems," Aerospace Corp., Aerospace Rept. No. TR-99(1670)-1, El Segundo, CA, July 1999.

[2]  Lai, S. T., "A Critical Overview on Spacecraft Charging Mitigation Methods," *IEEE Transactions on Plasma Science*, Vol. 31, No. 6, 2003, pp. 1118–1124.

[3]  Lai, S. T., "An Overview of Electron and Ion Beam Effects in Charging and Discharging of Spacecraft," *IEEE Transactions on Nuclear Sciences*, Vol. 36, No. 6, 1989, pp. 2027–2032.

[4]  Canfield, R. C., Hudson, H. S., and Pevtov, A. A., "Sigmoids as Precursors of Solar Eruptions," *IEEE Transactions on Plasma Science*, Vol. 28, No. 6, 2000, pp. 1786–1794.

[5]  Gussenhoven, M. S., and Mullen, E. G., "Geosynchronous Environment for Severe Spacecraft Charging," *Journal of Spacecraft and Rockets*, Vol. 20, No. 1, 1983, pp. 26–34.

[6]  Lai, S. T., Gussenhoven, M. S., and Cohen, H. A., "Range of Electron Energy Spectrum Responsible for Spacecraft Charging," AGU Spring Meeting, May 1982; also *EOS*, Vol. 63, No. 18, 1982, p. 421.

[7]  Lai, S. T., Gussenhoven, M. S., and Cohen, H. A., "The Concepts of Critical Temperature and Energy Cutoff of Ambient Electrons in High-Voltage Charging of Spacecrafts," *Spacecraft/Plasma Interactions and Their Influence on Field and Particle Measurements*, ESA SP-198, edited by A. Pedersen, D. Guyenne, and J. Hunt, European Space Agency, Noordwijk, The Netherlands, 1983, pp. 169–175.

[8]  Laframboise, J. G., Godard, R., and Kamitsuma, M., "Multiple Floating Potentials, Threshold Temperature Effects, and Barrier Effects in High Voltage Charging of Exposed Surfaces on Spacecraft," *Proceedings of International Symposium on Spacecraft Materials in Space Environment*, ESA SP-178, European Space Agency, Paris, 1982, pp. 269–275.

[9]  Laframboise, J. G., and Kamitsuma, M., "The Threshold Temperature Effect in High Voltage Spacecraft Charging," *Proceedings of Air Force Geophysics Workshop on Natural Charging of Large Space Structures in Near Earth Polar Orbit*, AFRL-TR-83-0046, ADA-134-894, Air Force Geophysics Lab., Hanscom AFB, MA, 1983, pp. 293–308.

[10] Lai, S. T., "Spacecraft Charging Thresholds in Single and Double Maxwellian Space Environments," *IEEE Transactions on Nuclear Science*, Vol. 38, No. 6, 1991, pp. 1629–1634.

[11] Lai, S. T., and Della-Rose, D., "Spacecraft Charging at Geosynchronous Altitudes; New Evidence of the Existence of Critical Temperature," *Journal of Spacecraft and Rockets*, Vol. 38, No. 6, 2001, pp. 922–928.

[12]   Hastings, D., and Garrett, H. B., *Spacecraft-Environment Interactions*, Cambridge Univ. Press, Cambridge, England, U.K., 1997.

[13]   Laframboise, J. G., and Parker, L., "Probe Design for Orbit-Limited Current Collection," *Physics of Fluids*, Vol. 16, No. 5, 1973, pp. 629–636.

[14]   Whipple, E. C., Jr., "Observation of Photoelectrons and Secondary Electrons Reflected from a Potential Barrier in the Vicinity of ATS-6," *Journal of Geophysical Research*, Vol. 81, No. 14, 1976, pp. 715–719.

[15]   Mizera, P. F., Koons, H. C., Schnause, E. R., Croley, D. R., Jr., Alan Kan, H. K., Leung, M. S., Stevens, N. J., Berkopec, F., Staakus, J., Lehn, W. L., and Nanewicz, J. E., "First Results of Material Charging in the Space Environment," *Applied Physics Letters*, Vol. 37, No. 3, 1980, pp. 276–277.

[16]   Olsen, R. C., McIlwain, C. E., and Whipple, E. G., "Observations of Differential Charging Effects on ATS-6," *Journal of Geophysical Research*, Vol. 86, No. A8, 1983, pp. 6809–6819.

[17]   Olsen, R. C., "Modification of Spacecraft Potentials by Plasma Emission," *Journal of Spacecraft and Rockets*, Vol. 18, No. 5, 1981, pp. 462–469.

[18]   Lai, S. T., Cohen, H. A., Aggson, T. L., and McNeil, W. J., "The Effect of Photoelectrons on Boom-Satellite Potential Differences During Electron Beam Ejections," *Journal of Geophysical Research*, Vol. 92, No. A11, 1987, pp. 12319–12325.

[19]   Lai, S. T., McNeil, W. J., and Aggson, T. L., "Spacecraft Charging During Ion Beam Emissions in Sunlight," AIAA Paper 90-0636, Jan. 1990.

[20]   Lai, S. T., "An Improved Langmuir Probe Formula for Modeling Satellite Interactions with near Geostationary Environment," *Journal of Geophysical Research*, Vol. 99, No. A1, 1994, pp. 459–468.

[21]   Garrett, H. B., and Spitale, G. C., "Magnetospheric Plasma Modeling (0–100 keV)," *Journal of Spacecraft and Rockets*, Vol. 22, No. 3, 1985, pp. 231–244.

[22]   Mizera, P. F., "A Summary of Spacecraft Charging Results," *Journal of Spacecraft and Rockets*, Vol. 20, No. 5, 1983, pp. 438–443.

[23]   Lai, S. T., and Tautz, M., "High-Level Spacecraft Charging in Eclipse at Geosynchronous Altitudes: A Statistical Study," *Journal of Geophysical Research*, Vol. 111, A09201, 2006, doi: 10.1029/2004JA010733.

[24]   Lai, S. T., "The Importance of Surface Conditions for Spacecraft Charging," AIAA Paper 2009-349, Jan. 2009.

[25]   Vasyliunas, V. M., "A Survey of Low-Energy Electrons in the Evening Sector of the Magnetosphere with Ogo 1 and Ogo 3," *Journal of Geophysical. Research*, Vol. 73, No. 9, 1968, pp. 2339–2385.

[26]   Meyer-Vernet, N., "How Does the Solar Wind Blow? A Simple Kinetic Model," *European Journal of Physics*, Vol. 20, 1999, pp. 167–176.

[27]   Harris, J. T., "Spacecraft Charging at Geosynchronous Altitudes: Current Balance and Critical Temperature in a non-Maxwellian Plasma," Master's Thesis, AFIT/GAP/ENP/03-05, Air Force Inst. of Technology, Wright-Patterson AFB, OH, March 2003.

[28]   Lai, S. T., "Dependence of Electron Flux on Electron Temperature in Spacecraft Charging," *Journal of Applied Physics*, Vol. 105, 2009, doi: 10.1063/1.3125517.

[29]   Tautz, M., and Lai, S. T., "Analytic Models for a Rapidly Spinning Spherical Satellite Charging in Sunlight," *Journal of Geophysical Research*, Vol. 110, A07, 2005, pp. 220–229, doi: 10.1029/2004JA010787.

[30]   Mandell, M., Katz, I., Schnuelle, G., Steen, P., and Roche, J., "The Decrease in
       Effective Photo- Currents due to Saddle Points in Electrostatic Potentials near
       Differentially Charged Spacecraft," *IEEE Transactions on Nuclear Science*, Vol. 26,
       No. 6, 1978, pp. 1313–1317.
[31]   Lai, S. T., Cooke, D., Dichter, B., Ray, K., Smith, A., and Holeman, E., "Bootstrap
       Charging on the DSCS Satellite," *Proceedings of the 7th Spacecraft Charging
       Technology Conference*, ESA-476, Noordwijk, The Netherlands, 2001, pp. 345–349;
       also Minor, J. (ed.), *Spacecraft Charging Conferences* [CD-ROM], NASA SEE/
       TP-2005-600, NASA Marshall Space Flight Center, Huntsville, AL, 2005.
[32]   Mandell, M. J., Cooke, D. L., Davis, V. A., Jongeward, G. A., Gardner, B. M., Hilmer,
       R. A., Ray, K. P., Lai, S. T., and Krause, L. H., "Modeling the Charging of
       Geosynchronous and Interplanetary Spacecraft Using Nascap-2K," *Advances in
       Space Research*, Vol. 36, 2005, pp. P2511–2515.
[33]   Krause, L., Cooke, D. L., Enloe, C. L., Font, G. I., Lai, S. T., McHarg, M. G., and Putz,
       V., "Bootstrap Surface Charging at GEO: Modeling and On-Orbit Observations from
       the DSCS-III B7 Satellite," *IEEE Transactions on Nuclear Science*, Vol. 54, No. 6,
       2007, pp. 1997–2003.
[34]   Tautz, M., and Lai, S. T., "Analytic Models for a Spherical Satellite Charging in
       Sunlight at Any Spin," *Annals of Geophysics*, Vol. 24, No. 10, 2006, pp. 2599–2610.
[35]   Lai, S. T., and Tautz, M., "Aspects of Spacecraft Charging in Sunlight," *IEEE
       Transactions in Plasma Sciences*, Vol. 34, No. 5, 2006, pp. 2053–2061.
[36]   "Special Issue on TSS-1R: Electrodynamic Tether-Ionospheric Interactions,"
       *Geophysical Research Letters*, Vol. 25, Nos. 4–5, 1998.
[37]   Enloe, C. L., Cooke, D. L., Pakula, W. A., Violet, M. D., Hardy, D. A., Chaplin, C. B.,
       Kirkwood, R. K., Tautz, M. F., Bonito, N., Roth, C., Courtney, G., Davis, V. A.,
       Mandell, M. J., Hastings, D. E., Shaw, G. B., Giffin, G., and Sega, R. M.,
       "High-Voltage Interactions in Plasma Wakes: Results from the Charging Hazards
       and Wake Studies (CHAWS) Flight Experiments," *Journal Geophysical Research*,
       Vol. 102, No. A1, 1997, pp. 425–433.
[38]   Stone, N. H., and Bonifazi, C., "The TSS-1R Mission: Overview and Scientific
       Context," *Geophysical Research Letters*, Vol. 25, No. 4, 1998, pp. 409–412.
[39]   Eliasson, L., Andre, M., Erikson, A., Norqvist, P., Norberg, O., Lundin, R., Holback,
       B., Koskinen, H., Borg, H., and Boehm, M., "Freja Observations of Heating and
       Precipitation of Positive Ions," *Geophysical Research Letters*, Vol. 21, No. 17, 1994,
       pp. 1911–1914.
[40]   Eriksson, A. I., and Wahlund, J-E., "Charging of the Freja Satellite in the Auroral
       Zone," *IEEE Transactions in Plasma Science*, Vol. 34, No. 5, 2006, pp. 2038–2045.
[41]   Katz, I., Parks, D. E., Mandell, M. J., Harvey, J. M., Wang, S. S., and Roche, J. C.,
       "NASACAP, A Three-Dimensional Charging Analyzer Program for Complex
       Spacecraft," *IEEE Transactions on Nuclear Science*, Vol. NS-24, No. 6, 1977,
       pp. 2276–2280.
[42]   Stannard, P. R., Katz, I., Mandell, M. J., Cassidy, J. J., Parks, D. E., Rotenberg, M., and
       Steen, P. G., "Analysis of the Charging of the SCATHA (P78-20) Satellite," NASA
       CR-165348, Dec.1981.
[43]   Mandell, M. J., Stannard, P. R., and Katz, I., "NASCAP Programmer's Reference
       Manual," S-Cubed, Rept. SSS-84-6638, La Jolla, CA, 1984.
[44]   Neergaard, L., Minow, J. I., McCollum, M., Katz, I., Mandell, M., and Davis, V.,
       "Comparison of the NASCAP/GEO, SEE Interactive Charging Handbook and
       NASCAP-2K.1 Spacecraft Charging Codes," *Proceedings of the 7th Spacecraft*

*Charging Technology Conference*, edited by R. A. Harris, ESTEC, Noordwijk, The Netherlands, 2001; also Minor, J. (ed.), *Spacecraft Charging Conferences* [CD-ROM], NASA SEE/TP-2005-600, NASA Marshall Space Flight Center, Huntsville, AL.

[45]    *EWB 5.0 User's Reference Manual*, NASA/Glenn Contract Report NAS#-25347, 31 July 1997.

[46]    Ferguson, D. C., Craven, P. D., Minow, J. I., and Wright, K. H., Jr., "A Theory of Rapid Charging Events on the International Space Station," AIAA Paper 2009-3523, June 2009.

[47]    Roussel, J. F., Rogier, F., Volpert, D., Forest, J., Rousseau, G., and Hilgers, A., "Spacecraft Plasma Interaction Software (SPIS): Numerical Solvers - Methods and Architecture," 9th Spacecraft Charging Technology Conference, April 2005; also Goka, T. (ed.), JAXA-SP-005-001E [CD-ROM], Japan Aerospace Exploration Agency, Tsukuba Space Center, Ibaraki, Japan.

[48]    Roussel, J.-F., Rogier, F., Dufour, G., Mateo-Velez, J.-C., Forest, J., Hilgers, A., Rodgers, D., Girard, L., and Payan, D., "SPIS Open-Source Code: Methods, Capabilities, Achievements, and Prospects," *IEEE Transactions on Plasma Science*, Vol. 36, No. 5, Part 2, 2008, pp. 2360–2368.

[49]    Muranaka, T., Hosoda, S., Kim, J.-H., Hatta, S., Ikeda, K., Hamanaga, T., Cho, M., Usui, H., Ueda, H. O., Koga, K., and Goka, T., "Development of Multi-Utility Spacecraft Charging Analysis Tool (MUSCAT)," *IEEE Transactions on Plasma Science*, Vol. 36, No. 5, 2008, pp. 2336–2349.

[50]    Hoffmann, R., Dennison, J. R., Thomson, C. D., and Albretsen, J., "Low-Fluence Electron Yields of Highly Insulating Materials," *IEEE Transactions on Plasma Science*, Vol. 36, No. 5, 2008, pp. 2238–2245.

[51]    Ferguson, D. C., Vayner, B. V., and Galofaro, J. T., "Solar Array Arcing in LEO; How Much Charge Is Discharged?," *Protection of Materials and Structures from Space Environment*, edited by J. L. Kleiman, Springer–Verlag, New York, 2006, pp. 9–19.

[52]    Ferguson, D. C., Vagner, B. V., Galofaro, J. T., and Hillard, G. B., "Arcing in LEO–Does the Whole Array Discharge?," AIAA Paper 2005-481, Jan. 2005.

[53]    Vayner, B., Galofaro, J., and Ferguson, D., "The Neutral Gas Desorption and Breakdown on a Metal-Dielectric Junction Immersed in a Plasma," AIAA Paper 2002-2244, May 2002.

[54]    Vayner, B., Galofaro, J., and Ferguson, D., "Interactions of High-Voltage Solar Arrays with Their Plasma Environment: Physical Processes," *Journal of Spacecraft and Rockets*, Vol. 41, No. 6, 2004, pp. 1042–1050.

[55]    Cho, M., Kim, J.-H., Hosoda, S., Nozaki, Y., Miura, T., and Iwata, T., "Electrostatic Discharge Ground Test of a Polar Orbit Solar Panel," *IEEE Transactions on Plasma Science*, Vol. 34, No. 5, 2006, pp. 2011–2034.

[56]    Masui, H., Toyoda, K., and Cho, M., "Electrostatic Discharge Plasma Propagation Speed on Solar Panel in Simulated Geosynchronous Environment," *IEEE Transactions on Plasma Science*, Vol. 36, No. 5, 2008, pp. 2387–2394.

[57]    Berthou, C., Boulanger, B., and Levy, L., "Plasma ESD Qualification Test Procedure of Alcatel Alenia Space Solar Array," *IEEE Transactions on Plasma Science*, Vol. 34, No. 5, 2006, pp. 2004–2010.

[58]    Violet, M. D., and Frederickson, A. R., "Spacecraft Anomalies on the CRRES Satellite Correlated with the Environment and Insulator Samples," *IEEE Transactions on Nuclear Science*, Vol. 40, No. 6, 1993, pp. 1512–1520.

[59]  Lai, S. T., Murad, E., and McNeil, W. J., "Hazards of Hypervelocity Impacts on Spacecraft," *Journal of Spacecraft and Rockets*, Vol. 39, No. 1, 2002, pp. 106–114.

[60]  Frederickson, A. R., Benson, C. E., and Cooke, E. M., "Gaseous Discharge Plasmas Produced by High-Energy Electron-Irradiated Insulators for Spacecraft," *IEEE Transactions in Plasma Science*, Vol. 28, No. 6, 2000, pp. 2037–2047.

[61]  Griseri, V., Perrin, C., Fukunaga, K., Maeno, T., Payan, D., Levy, L., and Laurent, C., "Space-Charge Detection and Behavior Analysis in Electron Irradiated Polymers," *IEEE Transactions in Plasma Science*, Vol. 34, No. 5, 2006, pp. 2185–2190.

[62]  Wrenn, G. L., and Smith, R. J. K., "Probability Factors Governing ESD Effects in Geosynchronous Orbit," *IEEE Transactions on Nuclear Science*, Vol. 43, No. 6, 1996, pp. 2783–2789.

[63]  Baker, D., and Lanzerotti, L. J., "Where Are the 'Killer Electrons' of the Declining Phase of Solar Cycle 23," *Space Weather*, Vol. 4, No. 7, 2006, p. S07001.

[64]  Wrenn, G. L., Chronology of 'Killer Electrons': Solar Cycles 22 and 23," *Journal of Atmospheric Solar-Terrestrial Physics*, Vol. 71, Nos. 10–11, 2009, pp. 1210–1218.

[65]  Rodgers, D. J., Levy, L., Latham, P. M., Ryden, K. A., Sorensen, J., and Wrenn, G. L., "Prediction of Internal Dielectric Charging Using the DICTAT Code," http://esa.spaceweather.net/spweather/workshops/proceedings _w1/posters/rogers16.pdf.

[66]  Sorensen, J., "An Engineering Specification of Internal Charging," http://conferences.esa.int/96a09/Abstracts35/paper/.

[67]  Santinin, G., Ivanchenko, V., Evans, H., Niemann, P., and Daly, E., "GRAS: A General-Purpose 3-D Modular Simulation Tool for Space Environment Effects Analysis," *IEEE Transactions on Nuclear Science*, Vol. 52, No. 6, 2006, pp. 2294–2299.

[68]  Mullen, E. G., Frederickson, A. R., Murphy, G. P., Ray, K. P., Holeman, E. G., Delorey, D. E., Robson, R., and Farar, M., "An Autonomous Charge Control System at Geosynchronous Altitude: Flight Results for Spacecraft Design Consideration," *IEEE Transactions on Nuclear Science*, Vol. 44, No. 6, 1977, pp. 2174–2187.

[69]  Torkar, K., Fazakerley, A., and Steiger, W., "Active Spacecraft Potential Control: Results from the Double Star Project," *IEEE Transactions on Plasma Sciences*, Vol. 34, No. 5, 2006, pp. 2046–2052.

[70]  Geis, M. W., Efremow, N. N., Krohn, K. E., Twichell, J. C., Lyszczarz, T. M., Kalish, R., Greer, J. A., and Tabat, M. D., "Theory and Experimental Results of a New Diamond Surface-Emission Cathode," *The Lincoln Lab Journal*, Vol. 10, No. 1, 1997, pp. 1–18.

[71]  Cooke, D. L., and Geis, M., "Introducing the Passive Anode Surface Emission Cathode," AIAA Paper 2002-4049, July 2002.

[72]  Iwata, M., Sumida, T., Igawa, H., Fujiwara, Y., Okumura, T., Khan, M. A. R., Toyoda, K., Cho, M., Hatta, S., Sato, T., and Fujita, T., "Development of Electron-Emitting Film for Surface Charging Mitigation: Observation, Endurance and Simulation," AIAA Paper 2009-560, Jan. 2009.

[73]  Ferguson, D. C., "Alternatives to the ISS Plasma Contacting Units," NASA/TM-2002-211488 Rept., AIAA Paper 2002-0934, Jan. 2002.

# CHAPTER 2

# Incoming and Outgoing Electrons

Shu T. Lai*
*U.S. Air Force Research Laboratory, Hanscom Air Force Base, Massachusetts*

## 2.1   INTRODUCTION

Having introduced spacecraft charging in a broad brush in Chapter 1, we now give more details on a centrally important topic: secondary and backscattered electrons. We will explain why this topic is centrally important and discuss various models for it.

## 2.2   FUNDAMENTAL PHYSICS OF SPACECRAFT CHARGING

When an object is placed in a plasma, whether in space or in the laboratory, the object is likely to receive more ambient electrons than ions. This is because electrons, being lighter than the ions, move much faster. Accumulation of excess electrons on the surface of an object generates a negative surface potential (voltage). This is why spacecraft often charge to negative voltages.

Spacecraft can charge to positive voltages, depending on the ambient electron temperature and electron energy distributions. They can also charge to positive voltages in sunlight or during artificial electron beam emissions from the spacecraft.

For a typical spacecraft surface capacitance, it takes milliseconds to reach equilibrium approximately. There is no exact equilibrium because the ambient plasma always fluctuates. For differentially charged surfaces with higher capacitances, it takes longer time to reach equilibrium. For most practical purposes, it is useful to consider spacecraft surface potentials at equilibrium. At equilibrium, the charging voltage depends on the balance of all incoming and outgoing currents. According to Kirchhoff's law, the sum of currents at every junction in a circuit is zero. The spacecraft surface is a junction. The current balance equation is of the form:

$$\sum_{k} J_k(\phi) = 0 \qquad (2.1)$$

*Senior Research Physicist, Space Vehicles Directorate, 0173.1 Associate Fellow AIAA.

where $J$ is the flux, which, by definition, is the current per unit area. The subscript $k$ labels the type of current, and $\phi$ is the surface potential.

## 2.3   SECONDARY AND BACKSCATTERED ELECTRONS

The general shape of a secondary electron yield (SEY) curve $\delta(E)$ is shown in Fig. 2.1. SEY is also called the secondary electron emission coefficient. For every (incoming) primary electron of energy $E$, there are $\delta$ (outgoing) secondary electrons generated from the surface. The energies of the secondary electrons are mostly at a few electron volts only. The $\delta(E)$ curve might cross unity twice or never, depending on the surface material. For incoming electrons with energies between $E_1$ and $E_2$ of the crossings, there are more electrons going out than coming in, implying positive voltage charging. Because the secondary electrons have only a few electron volts in energy, the positive voltage is low (about $1-2$ V) and usually ignored. Charging to high negative voltages is of concern because it can affect the instrument operations and scientific experiments as well as the power system onboard. Although electrons with energies in the range $E_1$ to $E_2$ are responsible for charging to positive voltages, those below $E_1$ and above $E_2$ are responsible for charging to negative voltages. If the incoming electrons are of various energies, these two camps of electrons compete with each other.

Backscattered electrons differ from secondary electrons in four aspects. First, a backscattered electron is practically the same electron (primary) that comes in. (We will not go deep into quantum properties here). Second, the energy of a backscattered electron is nearly the same as the primary electron. Third, the

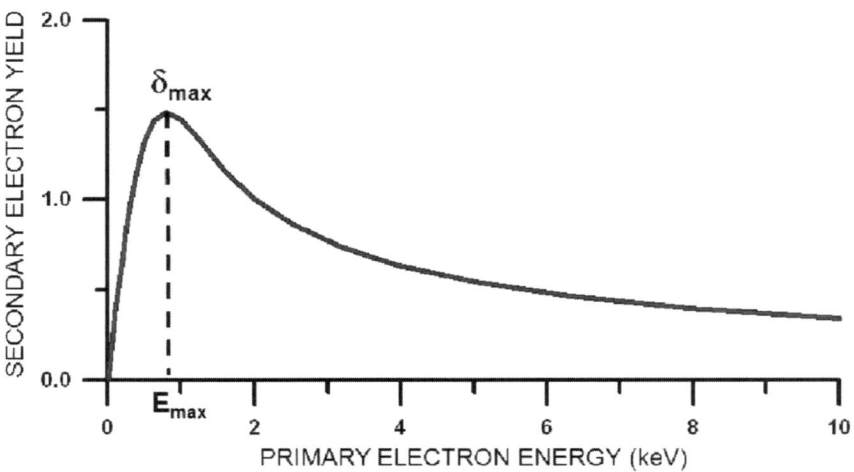

**FIG. 2.1   General shape of SEY $\delta(E)$.**

backscattered electron yield (BEY) $\eta(E)$ never exceeds unity. In the literature, BEY is also called electron reflection coefficient. Fourth, $\eta(E)$ is much smaller than unity and varies very slowly with $E$, except at energies $E$ below about 20 eV depending on the material.

## 2.4  CURRENT BALANCE

As mentioned in Sec. 2.2, electrons are faster than ions. The electron flux exceeds the ion flux by two orders of magnitude because of the mass ratio. The electron flux dominates. The onset of charging is determined by the current balance Eq. (2.2) of the electrons only. For incoming electrons having a distribution $f(E)$, the current balance between the incoming and outgoing electron fluxes is given by the equation:

$$\int_0^\infty dEEf(E) = \int_0^\infty dEEf(E)[\delta(E) + \eta(E)] \qquad (2.2)$$

For Maxwellian space plasmas, the solution of Eq. (2.2) yields a *critical temperature*, above which charging to negative voltages occurs [1-3]. Once charging sets in, the ambient ions are attracted towards the spacecraft. The charging voltage $\phi$ at equilibrium is given by the current balance between all incoming electrons, outgoing electrons, and incoming ions. Spacecraft charging is most important in the geosynchronous environment because the ambient plasma density is low and the plasma condition varies by large amplitudes during storms and substorms. In the geosynchronous environment, the Mott-Smith Langmuir model [4] is often a good approximation for describing the charge attraction. The current balance equation for negative charging voltage $\varphi(<0)$ at geosynchronous altitudes is of the form:

$$I_e(0)[1 - <\delta + \eta>]\exp\left(-\frac{q_e\phi}{kT_e}\right) = I_i(0)\left[1 - \frac{q_i\phi}{kT_i}\right]^\alpha \qquad (2.3)$$

where

$$<\delta + \eta> = \frac{\int_0^\infty dEEf(E)[\delta(E) + \eta(E)]}{\int_0^\infty dEEf(E)} \qquad (2.4)$$

In Eq. (2.3), $I_e(0)$ and $I_i(0)$ are respectively the ambient electron and ion fluxes at surface potential $\phi = 0$; $q_e(< 0)$ and $q_i(>0)$ are the electron and ion charges, respectively; $T_e$ and $T_i$ are the electron and ion plasma temperatures, respectively; $k$ is the Boltzmann constant; and $\alpha$ is the exponent of the attraction term. The exponent $\alpha = 1$ for a sphere, $\frac{1}{2}$ for an infinite cylinder, and 0 for an infinite plane. Using the functions $f(E)$, $\delta(E)$, and $\eta(E)$ as input, Eqs. (2.3) and (2.4) can be solved, numerically or analytically, for the spacecraft surface potential $\phi(< 0)$.

For arbitrary spacecraft geometries and arbitrary distributions $f(E)$ and arbitrary coefficients $\delta(E)$, and $\eta(E)$, one needs to use numerical methods to solve the current balance equation. NASCAP-2K [5], which is export controlled, is a very useful software for handling such numerical computations. Software for similar purposes and of similar capabilities are being developed independently at present by various teams in Japan, France, and Sweden.

## 2.5   OUTGOING ELECTRONS: SEY AND BEY FUNCTIONS

The Sternglass SEY $\delta(E)$ formula [6] has been widely used in spacecraft charging since the late 1970s for some 20 years. It is complicated, generally regarded as outdated, and no longer used.

The Sanders and Inouye SEY $\delta_S(E)$ formula [7] has an advantage. We will use a subscript to denote the first author. In this case, $s$ denotes Sanders. Here $\delta_S(E)$ is of exponential form, which is of great help for doing the integrations Eqs. (2.2) and (2.4) analytically.

$$\delta_S(E) = c\left[\exp\left(\frac{-E}{a}\right) - \exp\left(\frac{-E}{b}\right)\right] \tag{2.5}$$

where

$$a = 4.3E_m \qquad b = 0.367E_m \qquad c = 1.37\delta_m \tag{2.6}$$

where $E_m$ is the energy at which $\delta(E)$ is maximum and $\delta_m$ is the maximum value of $\delta(E)$.

The backscattering electron yield $\eta(E)$ formula of Prokopenko and Laframboise [8] has been often used for nearly three decades. The BSY formula needs to be updated. It is of the form

$$\eta(E) = A - B\exp(-CE) \tag{2.7}$$

Variation (Fig. 2.2) of the SEY and BEY functions affects the calculated results of not only critical temperature but also the equilibrium spacecraft potential.

The Katz et al. $\delta_K(E)$ formula [9] prescribes using a penetration range formula for primary electron energy $E$ above the second crossing energy $E_2$. However, for $E$ below $E_2$, the Katz et al. paper did not mention what to use. Presumably one can use the Sanders and Inouye formula for $E < E_2$. The Katz prescribed value of $E_2$, according to the paper, is "the value that extrapolates to unity." For gold, the prescribed value of $E_2$ is 4.6 keV. This value is nowhere near the second crossing of $\delta(E)$. A constant value for backscattering is prescribed by Katz et al. [9], but the value is higher than that given by Prokopenko and Laframboise [8], Eq. (2.7). For gold, the Katz et al. formulas [9] are as follows:

$$\delta_K = 0.52(E/E_2)^{1-p} \quad \text{for } E > E_2, \ E_2 = 4.6\,\text{keV}, \ p = 1.73 \tag{2.8}$$

$$\eta_K = 0.64 \tag{2.9}$$

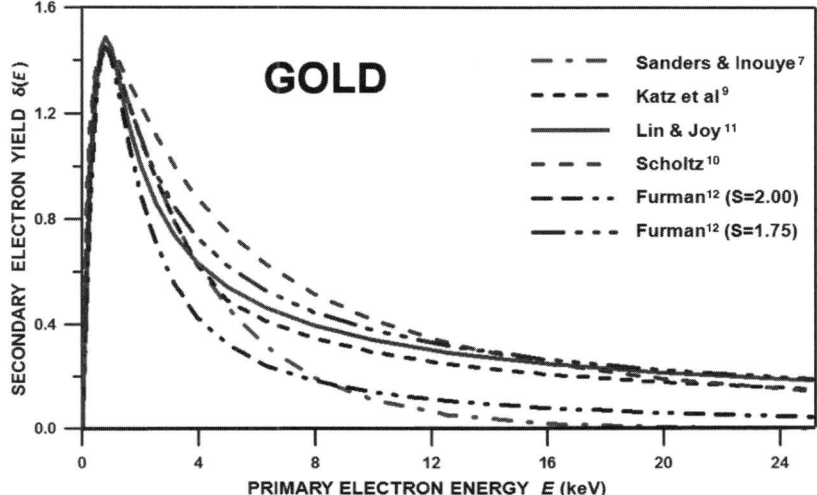

**FIG. 2.2   Secondary electron yield δ(E) functions for gold [16]. The functions published by different authors fall off differently beyond the peak.**

There are more recent $\delta(E)$ SEY formulas in the literature. The Scholtz SEY formula [10] is of the form

$$\delta_{SZ}(E) = \exp\left(-\log_e\{[E/E_M]^2/[2(1.6)^2]\}\right)\delta_M \tag{2.10}$$

The Lin and Joy SEY formula [11] is of the form

$$\delta_L = 1.28(E/E_M)^{-0.067}\{1 - \exp[-1.164(E/E_M)^{1.67}]\}\delta_M \tag{2.11}$$

In the Large Hadron Collider (LHC), where the world's most expensive experiments are to be conducted for hunting the Higg's boson, higher dimensions, and other physics frontiers, care is being paid to low-energy electron clouds generated from the walls that might affect high-energy particle trajectories. LHC has adopted the Furman SEY formula [12], which is different from all others by adding an empirical parameter $s$, characterizing the surface condition. The Furman formula [12] is of the form

$$\delta_F = \delta_M \frac{s(E/E_M)}{s - 1 + (E/E_M)^s} \tag{2.12}$$

where $s$ is the parameter to be determined by fitting the actual measurement, which varies depending on the surface condition, such as roughness, surface composition, contamination, etc.

The BEY $\eta(E)$ function also needs to be updated. As the primary electron energy $E$ decreases towards 0, the $\eta(E)$ function, Eq. (2.7), of Prokopenko and

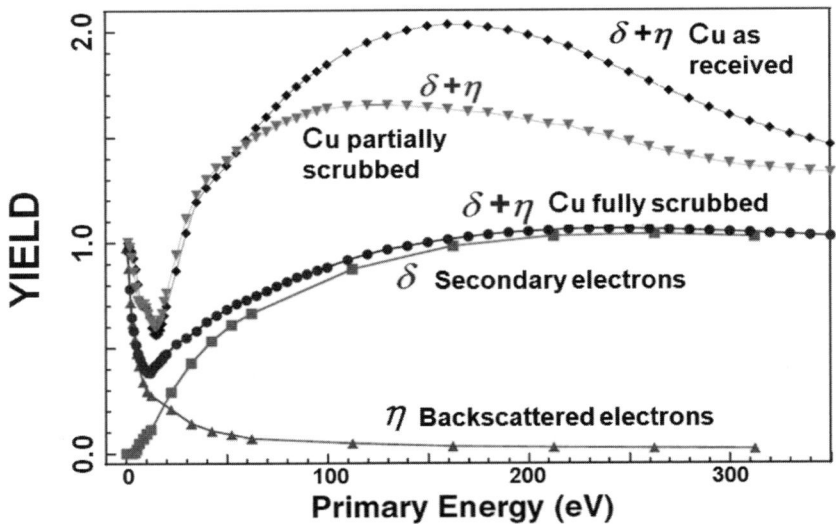

**FIG. 2.3    Measurements of SEY and BEY for the LHC accelerator tube walls at CERN [13, 14]. The SEY depends on the surface condition. The BEY rises to unity as the primary electron energy decreases to 0. (Reprinted with permission of Elsevier.)**

Laframboise [8] decreases monotonically to a small finite value. However, recent experiments [13, 14] on gold and other surface materials found the BEY function rising to unity as $E$ decreases to 0, instead of decreasing to a small value (Fig. 2.3). Jablonski and Jiricek [15] also found a similar feature for other materials.

Lai and Tautz [17] suggested a new $\eta$ formula by adding $\Delta\eta$ to that of Prokopenko and Laframboise [8]:

$$\eta \rightarrow \eta + \Delta\eta \tag{2.13}$$

$$\Delta\eta = (1 - A + B)\exp\left(-\frac{E}{E_0}\right) \tag{2.14}$$

In Eq. (2.14), the parameter $E_0$ has to be determined for each surface material. For gold, it is 0.05 keV [17].

## 2.6    RECOMMENDATIONS

Because SEY depends very much on the surface condition, it is suggested that a surface condition parameter $s$ be added to the BEY formula. The Furman formula [12] is a good candidate. Because BEY rises to unity as $E$ decreases to

0, the conventional $\eta$ formula of Prokopenko and Laframboise [8] should be revised. One suggestion is that an additional term $\Delta\eta$ be added to the commonly used $\eta$ formula.

We recommend that both SEY and BEY should be measured with actual samples before launch instead of taken merely from some materials encyclopedia. This is because the surface condition is important. Different surface conditions give different SEY and BEY results, which, in turn, give different values of critical temperature and spacecraft surface potential. It would be even better if changes in these coefficients could be monitored in space, which is hazardous at times, or simulated in parallel (with near real time) in the laboratory with the measurement data history of the space and surface conditions if possible.

## 2.7   INCOMING ELECTRONS: MAXWELLIAN AND KAPPA DISTRIBUTIONS

The Maxwellian velocity distribution of the ambient electrons can be written in terms of the electron energy $E$ by the relation: $E = (1/2)mv^2$, where $m$ is the electron mass and $v$ the electron velocity. As a result of the conversion from $v$ to $E$, the Maxwellian distribution $f(E)$ is of the form

$$f(E) = n\left(\frac{m}{2\pi k\,T}\right)^{1/2} \exp\left(\frac{-E}{kT}\right) \qquad (2.15)$$

Equation (2.15) is useful for Maxwellian spacecraft charging modeling [2] because its exponential form often enables analytical integrations to carry out. Using Eq. (2.15) and the SEY and BEY functions, one can calculate the critical temperature for the onset of spacecraft charging. If one also knows, in addition to the electron distribution, the ambient ion flux, one can calculate the spacecraft potential by using the current balance equation.

When the space plasma is highly perturbed by geomagnetic storms, for example, the electron and ion distributions often deviate from being Maxwellian. In such situations, the kappa distribution is often more suitable. The kappa $f_\kappa(E)$ is of the form

$$f_\kappa(E) = n\frac{\Gamma(\kappa+1)}{\Gamma(3/2)\Gamma(\kappa-1/2)}\left[\left(\kappa-\frac{3}{2}\right)T_\kappa\right]^{-\frac{3}{2}}\left[1+\frac{E}{(\kappa-3/2)T_\kappa}\right]^{-(\kappa+1)} \qquad (2.16)$$

where the kappa temperature $T_\kappa$ is related to the usual temperature $T$ as

$$T_\kappa = \frac{\kappa}{(\kappa-3/2)}\,T \qquad (2.17)$$

Note that

$$\frac{3}{2} < \kappa < \infty \qquad (2.18)$$

In the limit $\kappa \to \infty$, one recovers the Maxwellian distribution, Eq. (2.15). By using the kappa distribution, Eq (2.16), one can obtain the critical kappa temperature $T_\kappa^*$, from which one obtains the usual critical temperature $T^*$ by means of Eq. (2.17). Harris [18] examined the Los Alamos National Laboratory (LANL) geosynchronous spacecraft charging data and concluded that, at the onset of spacecraft charging, the electron temperature is not hot enough to render the distribution deviating substantially from being Maxwellian. At higher temperatures, the distribution deviates more.

## REFERENCES

[1]    Lai, S. T., "Spacecraft Charging Thresholds in Single and Double Maxwellian Space Environments," *IEEE Transactions on Nuclear Science*, Vol. 38, No. 6, 1991, pp. 1629–1634.

[2]    Lai, S. T., and Della-Rose, D., "Spacecraft Charging at Geosynchronous Altitudes; New Evidence of the Existence of Critical Temperature," *Journal of Spacecraft and Rockets*, Vol. 38, No. 6, 2001, pp. 922–928.

[3]    Lai, S. T., and Tautz, M., "High-Level Spacecraft Charging in Eclipse at Geosynchronous Altitudes : A Statistical Study," *Journal of Geophysical Research*, Vol. 111, No. A09201, 2006, doi: 10.1029/2004JA010733.

[4]    Mott-Smith, H. M., and Langmuir, I., "The Theory of Collectors in Gaseous Discharges," *Physical Review*, Vol. 28, No. 4, 1926, pp. 727–763.

[5]    Mandell, M. J., Cooke, D. L., Davis, V. A., Jongeward, G. A., Gardner, B. M., Hilmer, R. A., Ray, K. P., Lai, S. T., and Krause, L. H., "Modeling the Charging of Geosynchronous and Interplanetary Spacecraft Using Nascap-2K," *Advances in Space Research*, Vol. 36, 2005, pp. 2511–2515.

[6]    Sternglass, E. J., "Theory of Secondary Electron Emission," Westinghouse Res. Lab., Sci. Pap. 1772, Pittsburgh, PA, 1954.

[7]    Sanders, N. L., and Inouye, G. T., "Secondary Emission Effects on Spacecraft Charging: Energy Distribution Considerations," *Spacecraft Charging Technology 1978*, edited by R. C. Finke and C. P. Pike, NASA-2071, ADA-084626, AFGL, Hanscom AFB, MA, 1978, pp. 747–755.

[8]    Prokopenko, S. M., and Laframboise, J. G. L., "High Voltage Differential Charging of Geostationary Spacecraft," *Journal of Geophysical Research*, Vol. 85, No. A8, 1980, pp. 4125–4131.

[9]    Katz, I., Mandell, M., Jongeward, G., and Gussenhoven, M. S., "The Importance of Accurate Secondary Electron Yields in Modeling Spacecraft Charging," *Journal of Geophysical Research*, Vol. 91, No. A12, 1986, pp. 13,739–13,744.

[10]   Scholtz, J. J., Dijkkamp, D., and Schmitz, R. W. A., "Secondary Electron Properties," *Philips Journal of Research*, Vol. 50, No. 3–4, 1996, pp. 375–389.

[11]   Lin, Y., and Joy, D. G., "A New Examination of Secondary Electron Yield Data," *Surface Interface Analysis*, Vol. 37, 2005, pp. 895–900, doi: 10.1002/sia.2107.

[12]   Furman, M. A., "The Electron-Cloud Effects in the arcs of the LHC," LHC Project Rept., CERN, Geneva, 1998.

[13]   Cimino, R., Collins, I. R., Furman, M. A., Pivi, M., Ruggerio, F., Rumulo, G., and
       Zimmermann, F., "Can Low-Energy Electrons Affect High-Energy Physics
       Accelerators?," *Physical Review Letters*, Vol. 93, No. 1, 2004, pp. 14,801–14,804.

[14]   Cimino, R., "Surface Related Properties as an Essential Ingredient to e-cloud
       Simulations," *Nuclear Instruments and Methods in Physics Research, Section A:
       Accelerators, Spectrometers, Detectors and Associated Equipment*, Vol. 561, No. 2,
       June 2006, pp. 272–275.

[15]   Jablonski, A., and Jiricek, P., "Elastic Electron Backscattering from Surfaces at Low
       Energies," *Surface Interface Analysis*, Vol. 24, 1996, pp. 781–785.

[16]   Lai, S. T., "The Importance of Surface Conditions for Spacecraft Charging," *Journal
       of Spacecraft and Rockets*, Vol. 47, No. 4, 2010, pp. 634–638.

[17]   Lai, S. T., and Tautz, M., "On the Anti-Critical Temperature in Spacecraft Charging,"
       *Journal of Geophysical Research*, Vol. 113, No. A11211, 2008, doi: 10.1029/
       2008JA013161.

[18]   Harris, J. T., "Spacecraft Charging at Geosynchronous Altitudes: Current Balance
       and Critical Temperature in a Non-Maxwellian Plasma," M.S. Thesis, AFIT/GAP/
       ENP/03-05, Air Force Inst. of Technology, Wright-Patterson AFB, OH, March 2003.

# CHAPTER 3

# Spacecraft Charging, Arcing, and Sustained Arcs in Low Earth Orbit

## Dale C. Ferguson[*]
*U.S. Air Force Research Laboratory, Kirtland Air Force Base, New Mexico*

## G. Barry Hillard[†]
*NASA Glenn Research Center, Brookpark, Ohio*

## 3.1  INTRODUCTION

### 3.1.1  REASONS FOR USING HIGH-VOLTAGE SYSTEMS AND POSSIBILITY OF CHARGING IN LEO

High-voltage systems are used in space in order to save launch weight. First of all, for the same power level, higher voltages enable use of thinner wires (lighter cabling). This is because $P = IV$, and $V = IR$, so $P = I^2R$. If $I$ is decreased by use of higher $V$, then the wire resistance can be increased, that is, thinner wires can be used, with no increase in power loss due to cabling. In the case of the International Space Station (ISS), the decision to use a 160-V primary power system was based on the decreased cable mass possible. Of course, if one uses the same cable mass, higher voltages will enable higher efficiencies, as less power will be lost to resistance in the cables. For very large power systems, the decrease in cable mass can be substantial.

Secondly, some spacecraft functions require high voltages. For example, electric propulsion uses voltages from about 300 V (Hall thrusters) to about 1000 V (ion thrusters). For low-voltage power systems, conversion of substantial power to high voltages is required for these spacecraft functions to operate. The weight of the power conversion systems (PMAD) can be a substantial fraction of the total power system weight in these cases. It is more efficient, and can save weight, if the high-voltage functions can be directly powered from a high-voltage solar array, for instance. If the high-voltage function is electric propulsion, we call such a system a direct-drive electric propulsion system. Systems have been proposed that switch between parallel and series combinations of different strings of solar cells in order to facilitate occasional high-voltage requirements, but to enable housekeeping functions at a lower voltage.

---

[*]Lead, Spacecraft Charging Science and Technology.
[†]Group Lead for Spacecraft Charging and Arcing, 44106.

However, space power systems in low Earth orbit (LEO) can charge negatively up to 90% of their power system voltage, and this in some cases can lead to electrostatic discharge (arcing), and sometimes to sustained discharges, with varying degrees of serious consequences for their spacecraft. In this chapter, we discuss spacecraft charging, arcing, and sustained discharges in LEO.

### 3.1.2  ALTITUDE AND LATITUDE RANGE OF APPLICABILITY

The guidelines in this chapter are intended for space systems that spend the majority of their time at altitudes between 200 and 1000 km (usually known as LEO applications) and at latitudes between about $+$ and $-50$ deg, that is, space systems that do not (often) encounter the auroral ovals of electron streams, that do not encounter GEO (geosynchronous orbit) charging conditions, and that do not fly through the Van Allen belts. For the extreme radiation protection that is necessary for those orbits, exterior spacecraft charging will likely be a secondary concern.

### 3.1.3  OVERVIEW OF PLASMA INTERACTIONS

When energized conductors are exposed to plasma, positive surfaces collect electrons while negative surfaces collect ions. The Poisson equation governs charge movement. The Poisson equation is

$$\nabla^2 \phi = -4\pi\rho \tag{3.1}$$

where $\varphi$ is the potential and $\rho$ is the charge density. When the charge density is very low, as in GEO, Poisson's equation reduces to Laplace's equation.

Electrons, which are much lighter and more mobile than ions, are collected more easily. Surfaces, therefore, charge to whatever potential they must in order for the net current flow to be zero in equilibrium. A current loop forms. The potential that any given surface will achieve is very difficult to model and generally requires full-up testing in a plasma environment. The resulting interactions can be summarized as follows:

- Surfaces that are more negative than $\approx 100$ V or so with respect to their surroundings are subject to arcing. These arcs can be either plasma arcs or arcs to adjacent conductors. They are usually a momentary discharge of accumulated energy, lasting only milliseconds, but under some conditions can be sustained. The necessary conditions are for the current and voltage to be maintained above threshold values. Plasma arc thresholds are poorly known but can be as low as 55 V under some conditions.

- Surfaces that are more negative than $\approx 100$ V or so are subject to ion bombardment and sputtering. Because the dominant ion is atomic oxygen, care must be taken that chemical attack does not occur as well.

- Surfaces that are positive can easily collect sufficient electrons to present a measurable power drain to the system. Referred to as "parasitic current collection," this can result in a few percent power loss to the system.

- If the power system is negatively grounded, as is most commonly done, the entire vehicle can float negative with respect to the ionosphere. The system potential can get as negative with respect to the ionosphere as the entire power system voltage. For systems with very large areas of high-voltage surfaces, such as the ISS, this effect is large and requires a plasma contactor to mitigate. Experience with high-voltage experiments in the orbiter payload bay indicates that significant charging of the orbiter does not occur. Generally, the large area of exposed metal presented by the main engine bells collects sufficient ion current to balance at a low potential unless ion currents of more than about 30 mA are emitted, or unless the engine bells are in the orbiter wake. If a system is completely isolated from the orbiter body, plasma interactions will result in the system itself floating negative with respect to the orbiter. The magnitude of these effects is difficult to quantify in advance, but should be addressed in the design.

## 3.2   ENVIRONMENTS

### 3.2.1   AMBIENT ENVIRONMENT: NEUTRAL ATMOSPHERE

The dominant environment between 100 and 1000 km is the neutral atmosphere. In this essentially collisionless regime, the gases are in hydrostatic equilibrium. Below about 100 km, where the atmosphere is homogeneous, the composition is approximately 80% $N_2$ and 18% $O_2$ with traces of $NO_2$, Ar, and other gases. Above 100 km atomic oxygen, the result of photodissociation of molecular oxygen comes to dominate. Above about 800 km the atmosphere is largely atomic hydrogen. At a 500-km altitude, the neutral density varies from $2 \times 10^6$ to $3 \times 10^8 \, cm^{-3}$, depending on solar activity and position in the orbit. The kinetic temperature of the gas is usually between 500 and 2000 K, and the ambient pressure is in the range of $10^{-10}$ to $5 \times 10^{-8}$ torr.

The neutral gas environment has been well explored and quantified. Empirical models based on in situ neutral composition and satellite drag measurements have evolved over the years into reliable predictors of the average composition and thermal structure of the thermosphere. The most notable of these models are the Mass Spectrometer Incoherent Scatter MSIS-86 [1, 2] model based on in situ satellite observations of neutral concentrations and the MSFC version of the Jacchia model derived from satellite drag measurements [3, 4], as well as the U.S. Standard Atmosphere [5, 6]. These models provide good estimates of the thermosphere environment as functions of altitude, longitude, latitude, local time, magnetic activity, and solar activity and are continually updated, as new information becomes available.

## 3.2.2   PLASMA ENVIRONMENT

On the sunlit hemisphere of the Earth, ultraviolet (UV) and extreme-ultraviolet (EUV) radiation penetrates the atmosphere ionizing and exciting the molecules present. As this radiation penetrates, it is increasingly absorbed until, at 60 km altitude, it almost completely disappears from the solar spectrum. At the same time, however, the neutral density has been increasing, and as it does the ionization density increases. The result is a fine balance between increasing density and increasing absorption that leads to the formation of layers that is the mean structure that we call the ionosphere. A highly dynamic plasma, the ionosphere's properties vary with altitude, latitude, time of day, and sunspot cycle. Local geomagnetic disturbances can cause dramatic variations over hours to weeks that are difficult to predict. Despite these complications, the broad features of the ionosphere can be described with simple models.

The variability with latitude, known since the 1920s, is so dramatic that the ionosphere is conventionally divided into three distinct regions: high latitude, midlatitude, and low latitude. The high-latitude region, sometimes called the polar region, couples directly to the magnetospheric tail through the auroral magnetic field lines. With low plasma densities and high energies, this coupling subjects it to spectacular variations during geomagnetic substorms. The low-latitude region near the magnetic equator is subject to instabilities linked to changes in the magnetospheric ring current. The easiest region to understand is the midlatitude region, which most closely follows classical ionospheric models.

Variation with altitude is perhaps the most important parameter for the spacecraft designer. This pronounced vertical structure is not simply a matter of height variation but reflects basic physical processes that differ in the resulting regions. Three processes, in particular, are responsible: 1) the sun's energy is deposited at various heights because of the absorption characteristics of the atmosphere, 2) the physics of recombination depends on density and therefore on altitude, and 3) composition of the atmosphere changes with height.

The lower limit of the ionosphere is somewhat arbitrary because plasma production falls off continuously with decreasing height. Historically, the ionosphere has been assumed to begin at about 50 km from the surface, as this is the altitude where plasma density becomes sufficient to noticeably affect radio wave propagation. There is no distinct upper limit, but 2000 km is generally used for most practical applications.

Four layers describe the vertical structure of the ionosphere. In the order of increasing altitude and increasing plasma density, these are designated as D, E, F1, and F2 regions. Their properties are summarized in Table 3.1.

Beyond the peak in the F2 layer, electron density decreases monotonically out to several Earth-radii. For altitudes up to and including the F2 peak, thermal energies of the electrons and ions are in the range of 0.1 to 0.2 eV, corresponding to kinetic temperatures of 1200 to 2400 K. Temperature rises monotonically beyond this point reaching several thousand electron volts in geosynchronous orbits.

**TABLE 3.1   NOMINAL PROPERTIES OF IONOSPHERIC LAYERS**

| Region | Nominal Height of Peak, km | Plasma Density at Noon, $cm^{-3}$ | Plasma Density at Midnight, $cm^{-3}$ | Dominant Ion |
|--------|---------------------------|-----------------------------------|--------------------------------------|--------------|
| D  | 90  | $\sim 1.5 \times 10^4$ | Vanishes             | $O_2^-$ |
| E  | 110 | $\sim 1.5 \times 10^5$ | $\sim 1 \times 10^4$ | $O_2^+$ |
| F1 | 200 | $\sim 2.5 \times 10^5$ | Vanishes             | $O^+$   |
| F2 | 300 | $\sim 1.0 \times 10^6$ | $\sim 1.0 \times 10^5$ | $O^+$  |

The F2 layer is the most important for spacecraft operations. This is where ISS lives, where the shuttle orbiter and most LEO spacecraft fly, and where the Hubble orbits to photograph the universe. Its boundaries and electron density are highly variable with a general erratic behavior imposed on large daily, seasonal, and solar cycle variations.

Ionospheric plasma distributions within the F-region have been extensively explored since the advent of bottomside sounders, long before in situ satellite observations were made. As a result, the general morphology of the F-region and some of its more prominent individual features are well understood. Although there are detailed features such as localized troughs, localized heating, and short temporal variations that are difficult to model, the overall global structure of the ionosphere is now well understood, and excellent ionospheric models exist for estimating and quantifying plasma distributions. In particular, the global International Reference Ionosphere (for example, IRI-90) model provides estimates of plasma concentrations, composition and temperatures, under varying solar activity conditions.

## 3.2.3   SPACECRAFT-INDUCED ENVIRONMENT

Spacecraft-induced environments can take many forms: neutral gases, ionized gases (plasmas), condensable gases, particulates, radiation, etc. In many cases, these environments can overwhelm the natural environment and can lead to undesirable interactions. Next, we will treat these types of environments separately.

Cold-gas thrusters and RCS can significantly increase the localized neutral pressure. This can be dangerous when there are exposed high-voltage conductors, as Paschen discharges can occur (see Sec. 3.4.1). In general, if the local neutral pressure is more than a millitorr and less than about a few torr, high-voltage electrical breakdowns can occur. At a voltage of 3500 V, the TSS-1R tether leaked gas into its deployer control reel enclosures, and the elevated neutral pressure led to Paschen discharge and loss of the mission. On the SAMPIE shuttle payload bay experiment, a local gas vent had to be moved to prevent Paschen discharge.

Helium is the most dangerous neutral effluent gas, as it has the lowest Paschen breakdown minimum voltage.

Ionized gases can be emitted by plasma sources such as hollow cathode plasma contactors or from neutral gas sources at high positive potentials. Locally, the plasma density can be greater than the ambient plasma density, and similar plasma interactions can occur with high-voltage components. On ISS, the plasma contacting units (PCUs), when operating, produce a local xenon plasma of much greater density than ambient. It has been estimated that the invisible plasma ball produced is some 8 m in radius before its density decreases below the ambient plasma density in LEO. Arcing and current collection from such a plasma could occur in much the same way as with an ambient plasma, implying that solar arrays and other active sites should be kept out of induced plasma plumes.

Condensable gases are effluents that can condense out on cold components and contaminate their surfaces. Oil and water vapor are two major condensables that can influence the interactions of spacecraft surfaces. In vacuum chamber testing, oil has been shown to prevent snapover on surfaces when high positive voltages are used (see section 3.3.1.4.3). Many oils, however, cannot withstand the LEO atomic oxygen environment on ram-facing surfaces but can build up on wake surfaces. Water vapor released on the night side can condense on insulating surfaces of solar arrays, etc., and can participate in arcing when the arrays become active in sunlight. It has been shown in laboratory testing that solar arrays that have been thoroughly baked out (heated in a vacuum for seven days) lose the water vapor contamination that is important in low voltage (100–300 V) arcing [7]. In LEO, however, a cold cycle is about 1/3 of every orbit. Even very well baked-out systems can have recondensation from effluents evolved during the night side of the orbit. Thin layers of condensed contaminants can concentrate electric fields above high-voltage conductors, even to the point where they undergo dielectric breakdown.

Particulates can be emitted or shaken from surfaces, but can also result from arcing or sputtering from spacecraft surfaces. Particulates can transfer small amounts of charge from one surface to another, but their major effect is in changing the characteristics of the surfaces to which they adhere. For instance, an insulating particle on a conductor that is at a high potential can concentrate the electric field structure locally, possibly leading to a reduced arcing voltage threshold.

Radiation can embed electrons deep within dielectrics where they can build up for days, weeks, or months until the dielectric breaks down under the induced electric field. In the natural environment, this will mainly happen in the auroral zones, radiation belts, and above the South Atlantic Anomaly, but radiation produced on or within a spacecraft can be important regardless of orbital position. Satellites using radioactive power sources must be designed to ameliorate this "deep-dielectric" charging, which is different from the typical "surface" spacecraft charging.

## 3.3   PLASMA INTERACTIONS

Exposed high-voltage conductors that do not exhibit corona or breakdown in a neutral gas can readily do so if the environment contains a significant ionized component. Although a high-voltage surface, solar-cell interconnects, for example, can be exposed to the ionized space plasma by design, surfaces can also find themselves at high voltages because of current collection from the plasma. The resulting equilibrium potentials that are assumed by surfaces result in the following effects that are described in the sections that follow:

• *Floating potential shifts*: In equilibrium, some parts of the spacecraft can be charged to voltages near the maximum voltage appearing on the solar array.

• *Parasitic power drain*: Direct loss of power as a result of current collection. This can be several percent of total power.

• *Sputtering*: Surfaces that charge negative will attract ions that in turn will result in sputtering of the material.

• *Arcing*: Negative surfaces undergo arcing when some critical threshold is exceeded.

### 3.3.1   CURRENT COLLECTION

#### 3.3.1.1   CURRENT-BALANCE CONDITION

In the weakly ionized low-density plasma found in LEO, current collection is completely described by Poisson's equation [Eq. (3.1)]. Positive surfaces readily attract electrons while negative surfaces attract the much more massive positive ions only with great difficulty. Because in equilibrium net current collection must be zero, surfaces will charge to equalize the net current of each polarity.

To illustrate the basic effects, consider first a hypothetical experiment. Suppose two metal spheres a few feet in diameter are initially connected by a conductor and placed in LEO some distance apart. Because electrons are collected more easily than ions, both spheres will charge to the same potential, within a volt or two of plasma potential. Now suppose a high-voltage battery is placed between them with one sphere connected to the negative terminal and the other to the positive. On Earth, in air, such an arrangement would result in half of the battery voltage appearing on each sphere. But in LEO, highly mobile electrons stream to the positive sphere while the negative sphere struggles to collect the massive ions. Both experience and modeling indicate that approximately 90% of the battery voltage will appear on the negative sphere while only 10% will be on the positive one with respect to the plasma potential.

The implications of this are considerable and often expensive. In the case of ISS, for example, the power system consists of solar arrays wired in a series-parallel arrangement to give a 160-V system. Because the main structure of ISS is "grounded" to the negative end of the array string, the entire space station would "float" more than 140 V negative with respect to the ionosphere. Such potentials are beyond the dielectric strength of the anodized coatings on the ISS aluminum structure and would lead to arcing into the space plasma and eventual destruction of the ISS thermal control system. This prospect required the addition of an active plasma contactor, a xenon hollow cathode discharge unit, to effectively ground the space station to the ionosphere. As it turns out, the ISS solar arrays are unusual in that they are poor electron collectors because of their welded-through design. Atypically, the ISS early mission-build structure usually doesn't charge more than 20 V or so negative with respect to the surrounding plasma even without the plasma contactors operating. However, as more solar arrays were put up, the charging level on ISS increased dramatically, justifying the added expense of the plasma contactors.

For conducting surfaces that are covered with insulators, some elapsed time might be necessary for the steady-state potential situation to be reached. The surfaces will charge until no further charge collection is necessary in equilibrium, and this is tantamount to charging up a capacitor with plate separation equal to the insulator thickness. Ion charging times in LEO can be considerable for typical anodized aluminum thicknesses. It is estimated, for instance, that in the daytime ionosphere ISS surfaces take about 4 s to fully charge, whereas on the morning terminator where the ionospheric ion density is at its lowest, charging times of 40 s or more can occur.

### 3.3.1.2   SHEATH EFFECTS

A positively charged spherical electrode will collect electrons when inserted in a plasma. The volume called the sheath, in which the electrode influences electrons, is larger than the sphere. For low voltages, the sheath thickness will be nearly the same as the Debye length [see Eq. (3.4)]. Some electrons will orbit around the electrode and escape out of the sheath. The collected or trapped electrons are said to be orbit-limited and are affected in a complex manner by the radius of the electrode, the electrode voltages, and the temperature and density of the free electrons.

A solar array looks to the plasma like a large rod electrode (like the wires and interconnects that are in contact with the plasma) rather than a spherical probe and is also surrounded by a sheath. Power loss as a result of plasma leakage current will become significant above 100 V for positive electrodes and is discussed next. Above a threshold voltage, which differs due to array design, arcing can be observed between the electrodes.

### 3.3.1.3   CURRENT COLLECTION BY STRUCTURES

#### 3.3.1.3.1   Electron Collection

LEO spacecraft are traveling subsonically with respect to the electrons in the ambient plasma. That is, at the plasma temperatures in LEO, the ambient electrons are moving at speeds greatly in excess of the orbital velocity. Thus, electrons can be collected on any conducting surface (exposed to the undisturbed plasma, that is, not in the plasma wake) that is not charged more than a few electron temperatures negative. In general, electron collection is well described by probe theory. For example, see [8]. For large surfaces, collection is best described by thin sheath probe theory. For structures smaller than a few times the Debye length [see Eq. (3.4)], orbit-limited theory can be used. Electron current collected from a plasma can be described by the equation $I_e = J_0 A_s$, where $I_e$ is the electron current, $A_s$ is the effective surface area for electron collection (either the plasma sheath area or the area of a sphere with the limiting orbit radius), and $J_0$ is the electron thermal flux, given by

$$J_0 = (n/4)(8kT_e/\pi m_e)^{1/2} = 2.68 \times 10^{-12} \, nT_e^{1/2} \text{ Amps/cm}^2 \qquad (3.2)$$

where $n$ is the electron density per cubic centimeter and $T_e$ is the electron temperature in electron volts.

Electron current collection by wires is important in the case of electrodynamic tethers or when structures such as self-extending masts with wire braces are used. For instance, on ISS it was found that several square meters of electron collecting wires on the array masts were connected to ISS ground. The array wing that was positive with respect to the plasma because of $v \times B \cdot l$ effects (described next) acted as an electron collector and became essentially grounded to the surrounding plasma. This complicated measurements of the vehicle charging as a result of solar-cell electron collection.

An electrodynamic tether is a long wire orbiting in the Earth's magnetic field that uses the electric field generated by its motion, the so-called $v \times B \cdot l$ field (where $v$ is the velocity, $B$ is the magnetic field, and $l$ is the length of the tether or structure), to produce power or propulsion. This concept was proved on orbit by the PMG experiment, where both modes of operation were produced by emitting electrons (by means of plasma contactors) either at the top or bottom of a 500-m tether to produce power (electron emission at the bottom) or propulsion (electron emission at the top). The maximum $v \times B$ on a structure in LEO is about 1/3 V/m.

In the case of the TSS-1R tether, its 20-km length produced a maximum of about 3500-V potential between its most positive and negative ends because it wasn't oriented perfectly perpendicular to the velocity vector and the Earth's magnetic field. A satellite at its upper end collected electrons, and an electron gun at the lower end emitted electrons to complete the circuit. When the electron gun was not in operation, a large resistance prevented the shuttle from being biased thousands of volts negative of its surrounding plasma. However, there remained a large voltage between the tether lower end and the shuttle orbiter. This enormous bias eventually led to a continuous arc on the tether (see Sec. 3.3.2.2.2), which broke, freeing the satellite and ending the experiment. During the arc, the satellite collected over 1 A of electron current to keep the arc going. Probe theory [9] is usually used to calculate the total current collected by a wire with distributed potentials. However, before the break, TSS-1R demonstrated that a satellite at a high positive potential could collect an anomalously large electron current. See [10–12].

In the proposed MSFC tether experiment (ProSEDS [13]) the electrodynamic tether would have been a bare wire, collecting current along its length, rather than just at its ends. In this case, arc mitigation requires, for example, graded insulation at the tether ends to eliminate the so-called triple points where high electric fields can lead to arcing.

Electron collection in LEO is also affected by the vehicle plasma wake. Because orbiting LEO spacecraft are moving supersonically with respect to the ambient ions, there is a wake behind each spacecraft devoid of ions. The electrons that initially enter the wake build up a space charge that repels all other electrons, and so the wake can be considered essentially devoid of electrons, compared to the ambient plasma. For most bodies, then, the only part that can collect ambient electrons is the ram-facing side. The CHAWS experiment [14, 15] showed that a large body in LEO has a very deep wake, with a wake electron density of $10^{-4}$ of the ambient electron density or less, but with a temperature 10 times or so of the ambient, in agreement with earlier measurements by [16, 17] among others.

If a piece of conductive structure is surrounded by insulating material and is at a high positive potential relative to the ambient plasma, it might be subject to snapover (see Sec. 3.3.1.4.3). This has the effect of greatly increasing its effective electron surface area, and so it can collect an order of magnitude or more current than one would naively suspect. Insulating structure surfaces reach equilibrium potential with the LEO plasma of only a few volts negative and do not thereafter collect current.

### 3.3.1.3.2  Ion Collection

While electrons are collected from all directions in LEO, LEO spacecraft are moving supersonically with respect to the ions, and therefore ions are only collected by ram surfaces. In fact, because many conducting parts of a structure are far greater in dimension than the plasma sheath, the effective flux of ions to their surfaces is essentially equal to the ram flux of ions on their front-facing surfaces. That is, $F = n\,v$, where $n$ is the electron (and ion) number density and $v$ is the magnitude of the spacecraft velocity. If we let $A_{\mathrm{ram}}$ be the ram-facing conductor projected area, and if we let $I_i$ be the ion current and $q$ the ion charge,

$$I_i = q\,n\,v\,A_{\mathrm{ram}} \tag{3.3}$$

which for LEO circular orbit becomes $1.2 \times 10^{-13}\,n\,A_{\mathrm{ram}}$ amperes, and for a density of about $10^{12}/\mathrm{m}^3$, this gives a current of about $0.1\ \mathrm{A/m^2}$. This, then, is a convenient rule of thumb for LEO ion current, about $0.1\ \mathrm{A/m^2}$.

Notice that for most purposes, the collected ion current depends only on the electron (and ion) density, whereas the electron current depends on the electron temperature, as well. To first order, then, when there is a current balance condition determining the floating potential, only changes in the electron temperature will cause changes in the floating potential. Insulating ram surfaces will float at a potential such that the ram ion and thermal electron currents are equal, or only a few volts negative at the most.

### 3.3.1.4   CURRENT COLLECTION BY SOLAR ARRAYS

#### 3.3.1.4.1   Electron Collection

Electrons can be collected on positively charged cells of solar arrays by the cell interconnects, wiring traces, or cell edges. Solar-array electron collection is intimately related to parasitic power drain, which is treated later in this chapter. However, here we will talk in more general terms.

For arrays that have fully exposed interconnects, cell edges, or power traces, electron collection is similar to that of wires or small spheres of the same total collecting area as the exposed conductors. One significant difference is because many solar cells have insulating cover slides. Because solar arrays by definition generate a voltage across each string, some of the solar cells, interconnects, or wiring will be at very different voltages than other parts. If a solar-array string has 400 silicon solar cells in series, for instance, one end of the string will be about 200 V more positive than the other. The total electron current collected will be the integral of the collection of all of the cells at their respective potentials away from the plasma potential. This depends, of course, on what the system ground is, and what the floating potential of the system is. Wherever the system floats with respect to the ambient plasma, only the cells and traces with positive potentials will collect many electrons.

If the array's exposed conductors are partially hidden from the ambient plasma (such as being underneath overhanging cover slides or between closely spaced solar cells), the cover slides can change the electron collection greatly. It has been shown that a coverslide with an overhang at least as big as the cell-plus-adhesive thickness will block electron collection at the cell edge very effectively, cutting it by a few orders of magnitude. Also, cell edges on cells that are separated by less than about 32 mils have greatly reduced electron collection [18, 19]. One way of thinking about this reduced electron collection is that it becomes difficult or impossible for thermal electrons to "make the turn" to be collected at the cell edges. For such solar arrays, it is often the case that the lower the ambient electron temperature, the greater the electron collection, as more of the ambient electrons can make the turn. This is the case for the ISS arrays, where the greatest amount of electron collection, and thus the worst system charging, occurs when the ambient electron temperature is the lowest.

It is possible for the solar arrays to undergo snapover if they are at high enough positive potentials. See Sec. 3.3.1.4.3 for details. It is believed that snapover depends on the secondary electron emission characteristics of the solar-array insulators. Contamination and/or texturing by atomic oxygen can decrease snapover. In ground tests, oil contamination was seen to completely prevent snapover on some samples. If snapover does occur, it is possible for the solar array to have an effective electron collection area as great as its entire geometrical area, rather than the tiny fraction of the array area that is normally occupied by interconnects or cell edges.

The solar array itself can provide a wake to block its own electron collection. For a sun-pointing array in equatorial LEO, the electron collection will be a maximum near sunrise and will shut off about noon when the array goes into its own wake. Of course, at night when the plasma is not dense and the array is not generating voltage, the electron collection will be minimal. Thus, solar-array electron collection in LEO is only important, and can only lead to a great deal of system charging, for about 1/3 of each orbit (the morning side).

### 3.3.1.4.2   Ion Collection

Snapover does not occur for ions, and the ion collection for solar arrays is almost always a linear function of negative voltage. Again, the total array collection is the integrated value of all negative cells at their respective potentials away from the ambient plasma, but for most solar arrays this collection is small compared to ion collection from the structure. In the case of ISS, for example, [20] could completely ignore solar-array ion collection in modeling the ISS floating potential. When the array is in its wake, ion collection is further reduced.

### 3.3.1.4.3   Snapover

The phenomenon of snapover was observed in the early 1980s when power system designers first began experiments with high-voltage arrays. Although broadly understood, many details of the process are controversial and remain an active area of research.

Suppose a flat conducting plate is covered with an insulator and that in this insulation there is a pinhole. If the plate is biased by a power supply and placed in plasma, it will collect current. For low voltages, current collection will be linear with bias voltage. Although the remaining surface cannot collect charge, it nevertheless is the source of an increasing electric field. This field results in ion bombardment of the insulator and secondary electron emission. The result is a rapidly growing sheath that collects charge and funnels it effectively to the pinhole. What is observed then is this: As voltage is increased from zero, current is collected linearly. At some point, current collection increases exponentially and finally saturates at a current level that is approximately the same as if the entire plate were conducting. On a solar array, the interconnects, wire traces, or cell edges act like pinholes—they are the conductors to which the current is funneled. The solar-cell substrate and/or cover slides act like the insulator in the preceding example: they are the dielectric that furnishes the secondary electrons and acts as a current-collecting plate.

The phenomenon is quite striking with conventional solar-array designs and is easily observed in plasma test chambers. Here we have solar cells that are covered by insulating cover slides connected to each other by small, exposed metallic interconnects. At low voltages the interconnects collect current roughly linearly with voltage. At around 150–200 V the onset of snapover can be observed, and by about 600 V the array is fully "snapped over."

Avoiding snapover has become a major design issue. Strategies include insulating all surfaces, where practical, and choosing insulators with low secondary electron emission yields. Although simply insulating all conducting surfaces provides initial protection, cracks or pinholes are difficult to avoid when materials must withstand years of exposure to harsh space conditions. Pinholes in high-voltage insulation usually expand as the large current density funneled through them destroys additional material. On the other hand, experience has shown that cracks or pinholes, if much smaller than the Debye length in the plasma, do not snap over. [Here $\lambda_D = 743(T_e/n)^{1/2}$, in centimeters, where $T_e$ is the electron temperature in electron volts and $n$ is the electron density in negative cubic centimeters. See Eq. (3.4)]. For LEO conditions, $\lambda_D$ can be as small as 0.1 cm.

As an example of a snapover-like effect on real solar arrays, we consider the data in Fig. 3.1. The Advanced Photovoltaic Solar Array (APSA) was a very lightweight design proposed for widespread use in the early 1990s. Originally designed for deployment in GEO, the blanket material was carbon-loaded Kapton®, which had sufficient conductivity to avoid differential charging that is a common problem in that environment. Proposals to adapt APSA technology to LEO recognized that atomic oxygen would destroy the blanket material within a matter of days. The LEO prototype was therefore designed with a blanket of germanium-coated Kapton®, which would be resistant to atomic-oxygen attack. This material is not as conducting as carbon but is still a weak conductor.

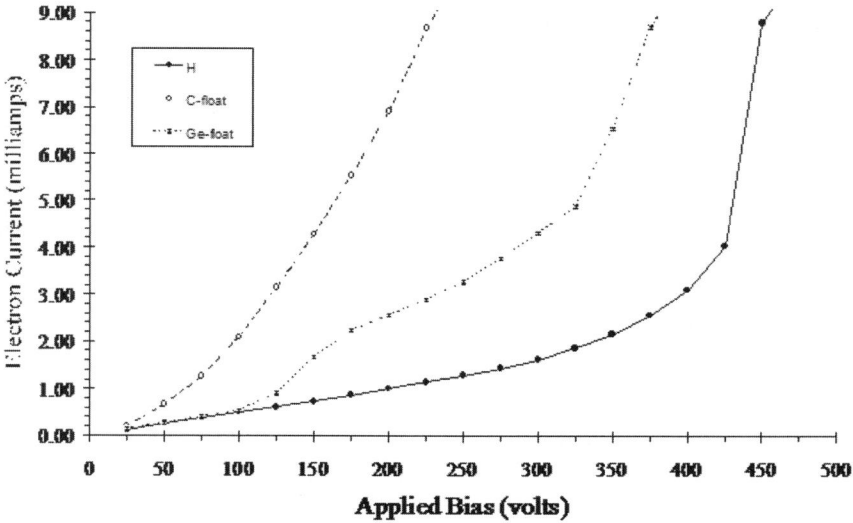

FIG. 3.1   Electron current vs bias for three solar-array blanket materials.

Three sample coupons that were as identical as possible except for the blanket material were constructed. One was made from uncoated Kapton®, whereas the other two had blankets coated with carbon and germanium, respectively. They were tested in a space simulation chamber for current collection as a function of applied bias voltage. As the results show, the highly insulating Kapton®-H, shown by the curve designated "H," collected current linearly until around 300 V. Current rose rapidly until about 400 V when it became exponential, the signature of snapover. The weakly conducting germanium-coated blanket collected linearly only until about 125 V when it began its rapid rise while the much more conducting carbon blanket collected exponentially almost from the beginning. These experiments showed that the blanket itself could become involved in the snapover process and pointed to the critical need to test all proposed array coatings for plasma effects (see [21]). That is, with conductive blankets, the inherent conductivity can substitute for the secondary electron-induced conductivity to give snapover even at low voltages.

### 3.3.1.4.4  Parasitic Power Drain

Current collection from solar arrays or other conducting surfaces not only poses the threat of damage to the surfaces involved, but also can reach levels that result in a significant loss of power. There have been many efforts over the years to use the basic equations of plasma physics to estimate the magnitude of this loss, and we will present one of them as illustrative of the effect.

The high-voltage solar-cell array for a high-power satellite looks more like a sheet electrode than like a spherical probe. K. L. Kennerud developed a method of analyzing the leakage current from such arrays [22] based on fundamental equations developed by I. Langmuir. Kennerud's technique converts the linear array into a sphere having the same area, and then he calculates the radius of the electron sheath surrounding the array. His experiments with small positively charged solar-cell panels correlated well with his predictions.

His results are shown in Table 3.2 and can be used to understand how the effect scales with altitude for the hypothetical solar array that he used.

Such rough calculations fail when the geometry becomes more complex. In particular, solar arrays with hidden interconnects, such as the ISS arrays, can collect current very differently than one with exposed interconnects. The ISS solar arrays, counter to intuition, collect more current at low electron temperatures than at high electron temperatures. Models have shown that this is caused by an electric field barrier to high-energy electrons. However, modeling electron collection by using spheres of equivalent "effective" area is very useful and is incorporated in computer codes such as EWB, for instance. Modern computer codes, such as the NASCAP series described in the following, will provide accurate estimates of parasitic power loss for any geometry. At high positive potentials snapover can make a solar array appear to be completely conductive. In addition, if a glow discharge caused by neutral gas ionization occurs on the array, the current collected can shoot up to tremendous levels [23, 24]. Finally, electric

### TABLE 3.2    LEAKAGE CURRENT FROM POSITIVELY CHARGED SOLAR ARRAYS

| Array Altitude, km | Electron Density, $N_e$, cm$^{-3}$ | Electron Temperature, K | Leakage Current | | Power Loss, % of Generated |
|---|---|---|---|---|---|
| | | | nA/cm$^2$ | A per 1500 V string[a] | |
| 500 | $6 \times 10^5$ | 3,000 | 824.5 | 0.8494 | 7.72 |
| 700 | $2 \times 10^5$ | 3,000 | 274.8 | 0.2831 | 2.57 |
| 1,000 | $7 \times 10^4$ | 3,000 | 96.19 | 0.0990 | 0.90 |
| 2,000 | $2 \times 10^4$ | 3,200 | 28.38 | 0.0292 | 0.265 |
| 30,000 | $1 \times 10^2$ | 13,600 | 0.29 | 0.0003 | 0 |

a   The string is 0.404 m by 255 m, with an area of 103.02 m$^2$.

propulsion thrusters or plasma contactors, if placed in the vicinity of solar arrays, can short circuit the plasma collection circuit and constitute a significant drain on the system power supply.

### 3.3.1.5   CURRENT COLLECTION AT HIGH FREQUENCIES

In general, little work has been done on plasma effects involving high-frequency power systems. Although significant new effects are not expected, most parameters of interest, such as corona inception and extinction voltages, are expected to exhibit frequency dependence. One effect did emerge in the early 1990s concerning insulated conductors energized with 20-kHz ac that were exposed to LEO plasma conditions. This work was underway because Space Station *Freedom* was originally designed to use such a power system [25]. Research was suspended when the space station was reconfigured to use dc power.

If a conductor energized with low-frequency ac is placed in LEO plasma, electrons are attracted to the insulating surface during the positive part of the cycle. These electrons "stick" to the material with a characteristic energy and are not repelled when the polarity changes to negative. Ions, however, are attracted during the negative part of the cycle and neutralize the electron charge for no net effect. At high frequencies this neutralization process does not occur. Highly mobile electrons are still attracted during the positive part of the cycle, but ions, because of the much larger mass, cannot respond to the rapidly changing field. The outer surface therefore charges to a negative potential close to the peak voltage on the power system waveform and remains charged.

Although ions cannot respond to the rapidly changing voltage waveform, they do respond to the buildup of negative charge on the surface. The resulting ion flux results in equilibrium where the surface is charged, as a rule of thumb, to about 90% of the peak voltage level used in the system. For a high-voltage system, ions will easily acquire sufficient energy to sputter material from the insulation.

Such charging might have a number of other implications that could include an arcing hazard, depending on where such surfaces are located with respect to other conductors.

### 3.3.1.6  WAKE EFFECTS

Because a LEO spacecraft is supersonic with respect to the ions it flies through, a wake, essentially devoid of plasma particles of both signs, will form behind it. In LEO, the ambient ions are traveling at a thermal speed of about $9.79 \times 10^5$ $(T_i/m_i)^{1/2}$ cm/s, where $T_i$ is the ion temperature in electron volts and $m_i$ is the ion mass in atomic mass units. For a $T_i$ of 0.2 eV (typical) and $m_i = 16$ (atomic oxygen), this gives an ion speed of about $1.1 \times 10^5$ cm/s, and a Mach ratio of about 7 for LEO orbit. Thus, the wake of a large body will extend as a cone about seven times as long as it is wide. In this region (a sort of umbra), ion and electron densities will be severely depressed, and the remaining plasma will be at a high temperature (perhaps 10 times that of the ambient plasma). In a surrounding region (a kind of penumbra), bounded by the shock wave, the plasma will be disturbed, but it is believed that the major effect will be hotter electrons than ambient. Beyond the penumbra, the plasma will be normal. For example, see [26]. Measured details of wake structure can be found in [16, 17].

Instruments to measure plasma parameters in LEO should be placed beyond the plasma sheath surrounding the structure (normally a distance of 0.3 to 0.6 m will suffice) and outside the wake of any structural element. In the case of the floating potential probe (FPP) on ISS, a compromise position was chosen that placed FPP outside the umbra of any structural element and on a pole to place it outside the plasma sheath, but it could not be placed out of the penumbra of some structural elements. Resulting plasma temperatures measured by FPP are considered to be higher than ambient temperatures, but the plasma densities seem reasonable. For instruments in such suboptimal placements, calibration must be done to convert measured parameters into ambient values, and such work is now proceeding with FPP. For a detailed discussion of wakes of large and small bodies orbiting in LEO, see [27]. In a practical document, we cannot go into great scientific detail about wake structure, but invite the reader to a literature search under the name of N. H. Stone, who had devoted much of his life to researching this topic.

### 3.3.2  ARCING

#### 3.3.2.1  SOLAR-ARRAY ARCING

##### 3.3.2.1.1  Background

Solar arrays have proven to be the major source of reliable long-term electric power for both manned and unmanned orbital spacecraft. In the early days of spaceflight, a few satellites used energy sources other than solar arrays and batteries, such as fuel cells or thermoelectric prime power sources. By 1970, because of extended mission times as well as increased power requirements, the majority of spacecraft primary power systems used solar arrays and rechargeable batteries to supply the required 28 V. The choice of 28 V for the main bus voltage was made to take advantage of long-existing standards and practices within the aircraft industry.

Plasma interactions at 28 V have not been generally considered a degradation factor of consequence. The only noted exceptions to their benign nature have occurred under extreme environmental conditions, especially during geomagnetic substorms for spacecraft operating at high inclinations. For low-inclination spacecraft, that is, those that completely avoid the auroral oval, 28-volt systems have not been observed to arc.

As the power requirements for spacecraft increased, however, high-voltage solar arrays were baselined to minimize total mass and increase power production efficiency. With the advent of 100-V systems in the late 1980s, arcing began to be observed on a number of spacecraft.

Solar-array arcs are generally characterized by several parameters:

- *Breakdown voltage*: The voltage required to initiate an arc depends on the plasma flux density, the system bias voltage, insulation, and construction and arrangement of the solar cells and solar-cell strings. Voltage breakdown for a well-designed solar array can initiate as low as 75 V (negative biased) for spacecraft operating in a LEO plasma environment. Reference [28] showed that arc thresholds of less than about 300 V are invariably caused by surface contamination with water and/or other contaminants.

- *Temporal profile*: The time from initiation to maximum current can be from a fraction of a microsecond to seconds, depending on the power source and the circuit impedance. The total duration of an arc can be from microseconds to indefinitely sustained.

- *Current profile*: The arc current can be as large as 100 to 1000 A depending on the capacitance of the solar array (see Fig. 3.2, from Snyder [29]).

Although many different taxonomies have been proposed for classifying arcs based on combinations of the preceding properties, these have generally been the work of physicists and have been designed to clarify issues for further research. For the design engineer concerned with risk mitigation, we will use a simpler scheme that assigns arcs to only two categories:

1. *Fast transients*: These are the most common solar-array arcs and are characterized by rapid rise time followed by extinction in a time that is several times the rise time. The critical parameter is that the energy involved is stored in whatever capacitance is available. The available capacitance can vary from a single array string to the entire spacecraft depending on design. These arcs give rise to electromagnetic interference (EMI) but otherwise are not generally associated with significant permanent damage on small spacecraft. On ISS

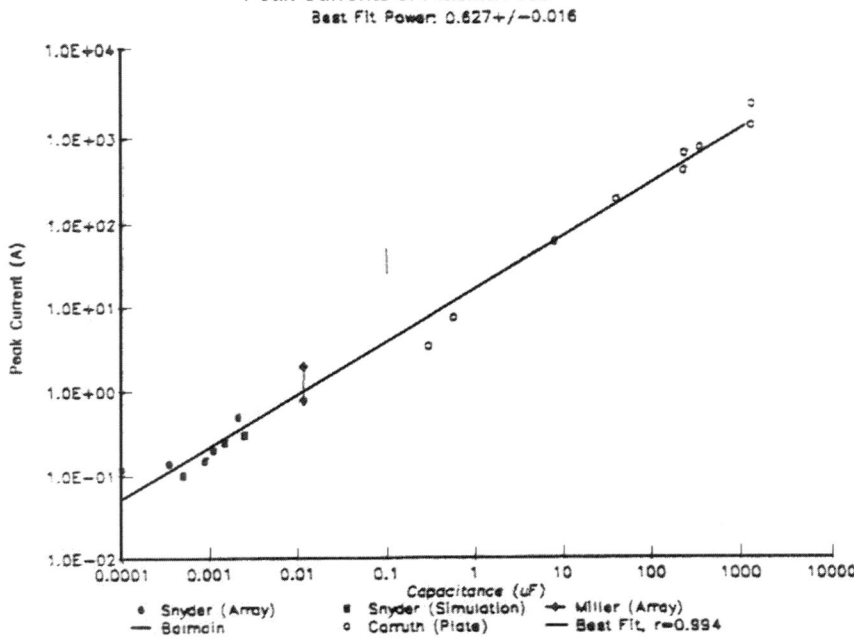

**FIG. 3.2    Peak arc current vs capacitance.**

and other high-power systems, however, the energy stored in the capacitance electrically connected to the arc site could cause significant damage to a solar cell or power trace. Of course, repeated arcs at the same arc site can lead to degradation and failure even if the individual arcs are not very energetic.

2. *Sustained arcs (continuous arcs)*: These are the events that have been attributed with the destruction of on-orbit solar arrays. Generally, the process begins with a fast transient (a so-called "trigger arc"). Under some conditions, the transient develops into an arc that is fed directly by the entire array, effectively becoming a short circuit. Such events invariably involve large quantities of energy and can be severely damaging to cells, inter-connects, or power traces.

Each of these will be discussed in more detail in sections to follow. Because all events begin as a fast transient and most do not evolve beyond this phase, this type of arc has been the object of most research in solar-array arcing. The more destructive continuous arc has only been observed in the past few years as power levels have increased (causing higher and higher string voltages to be used) and as the drive to ever more compact string layouts has resulted in some unfortunate design choices. The sections that follow are therefore organized around the fast transient event. We will turn to the continuous arc in the final section of this topic with a summary of what is known at this time.

### 3.3.2.1.2  Initiation Mechanism

As one might expect, the initiation of a solar-array arc depends on the presence of a strong local electric field. Frequently, the source is an exposed interconnect, which, depending on its location in the string, can be at high potential.

Most problematic are arcs that initiate at triple points. A triple point is a point in space where insulator, conductor, and plasma all meet. For a solar cell operating in LEO, this is usually the solar-cell interconnect, but it can also be the edge of the solar cell (near the substrate or the coverslide). It has been shown that arcing on solar arrays at voltages less than about 1000 V is always intermediated by the presence of a plasma. Identical samples to those that arced at 100 V in a plasma have been shown to withstand 1000-V bias in a pure vacuum.

Arcs have been observed at relatively low potentials (as low as 75 V) when conductor surfaces are biased negative near insulator surfaces in the presence of a plasma. Arc rate is strongly dependent on plasma density and on coverslide temperature, which affects the surface conductivity. It can range from intermittent (on a scale of minutes and perhaps hours or longer) to several per second. Arc currents observed in ground tests are on the order of an ampere and can last several microseconds. These characteristics depend on the capacitance to space, increasing with increasing capacitance. These arcs are usually associated with solar-cell array interconnects, but have also been observed on biased conductor surfaces covered with dielectric strips. They are likely to be of concern whenever conducting surfaces at negative potentials with respect to plasma abut insulating surfaces.

Several mechanisms are proposed for initiation of the arcs. Because much higher voltages are required to initiate arcs in a pure vacuum than in a plasma, the plasma arc must not be a so-called vacuum arc, but is initiated at much lower electric field strengths. One favored mechanism proposes that a thin layer of relatively insulating film develops on the conductor. High electric fields develop across the film as a result of ion collection on the exposed face. The resulting electric field across the film causes electron emission from the conductor through the film into the plasma [30]. A second, though perhaps related mechanism, assumes that the high electric fields at the edge of the dielectric cause propagation of secondary electrons to the dielectric surface from near the conductor-dielectric-vacuum interface. Also, sufficiently intense electric fields can develop locally at the tips of structure built on the conductor surface as a result of the mobility of surface atoms driven by the electric field resulting from the presence of the nearby dielectric surface. However, this "structure" related arcing requires thin whiskers that have not been seen on realistic samples.

Finally, gas desorbed from dielectric surfaces by electron impact can become ionized and serve as an ideal current path for the full-fledged arc.

At this time, no complete theories exist for the arc mechanism on solar-cell arrays in a plasma. All require inclusion of an empirical factor to produce the observed low-arcing voltage thresholds at triple points. Experimental evidence indicates that an electron emission mechanism plays an important role in producing the arcs. A preliminary theory that relates electron emission to the charging of a "dirty" layer on metal surfaces and the electric fields near an insulator-conductor-insulator surface configuration has been advanced. This theory accounts for some of the experimental observations.

An electron emission mechanism for solar-array arcing is consistent with several experimental observations. Kennerud [22] observed that the apparent ion collection of a solar-cell array was enhanced by an order of magnitude prior to arcing. This could be accounted for either by electron emission, or by an increase in ion density of the plasma. Snyder and Tyree [31] observed this emission as an increase in electron current collected by sensors in the tank with the solar array. They also noticed that these currents did not cease when the plasma generator was turned off. Arcing could still occur with no plasma in the tank as long as these emission currents were detected. Snyder [32] also noticed that arcs did not take place in a very low-density plasma ($10^2$ cm$^{-3}$).

The occurrence of arcs can be predicted from the potential of the solar-array cover slides relative to the plasma. In a very low-density plasma, even at relatively high bias voltages, the cover slides remained near plasma ground, and no arcs occurred. At higher plasma densities, the cover-slide potentials became several tens of volts more negative than plasma ground. When this condition existed, arcs occurred. Electrons from the plasma do not have enough energy to pass through the energy barrier set up by the biased interconnects and reach the insulator surfaces [33]. Electrons emitted from the interconnects of the array cause the cover slides to charge negatively relative to the plasma. These observations indicate that electron emission is necessary before the current pulse of the arcs can occur. Galofaro et al. [34] have shown that an arc is always preceded by a nanosecond burst of electrons from the arc site. This burst can also ignite arcs on nearby surfaces.

Jongeward et al. [30] proposed an arc mechanism model to account for this emission. The negatively biased interconnects tend to collect positive ions from the plasma. A layer of relatively high-resistance material several angstroms thick can collect a sufficiently high surface density of positive ions to permit field emission of electrons from the region. This mechanism was first proposed to account for enhanced secondary electron yields from oxide films [35]. Electrons emitted from this site are accelerated by the electric field between the cell or interconnect and the coverglass surface and strike the coverglass edge, which then emits secondary electrons in a cascade. Adsorbed gases are desorbed by electron impact. Ionization of these desorbed gases produces a dense plasma that is necessary for large currents to flow [36]. Some inferences that are consistent with the experimental observations can be made. There must be enough ion flux to the interconnect to maintain a high surface charge on the high-resistance layer. The metal-insulator geometry provides a focusing effect that increases the ion flux to the interconnect and maintains the surface charge density. Field emission accounts for the relatively steady emission, which probably represents a metastable situation. The solar-array arcs arise when this stability breaks down producing increased electron emission.

This model predicts the time duration and current of the arcs to within almost a factor of two. Progress is also being made in predicting arc rates using this model. For instance, de la Cruz et al. [37] were successful in modeling the arc rates and thresholds seen in the SAMPIE experiment. The importance of adsorbed contaminants has been experimentally verified by Vayner et al. [7] ( subsequent).

Brandhorst and Best [38] have shown that solar-array arcs can be initiated in the laboratory by simulated micrometeoroid strikes.

### 3.3.2.1.3 Arcing Threshold

In an attempt to consolidate all known arcing information on solar arrays, Ferguson [39] analyzed the arcing data from the PIX II array and compared it to other ground and flight data (see Fig. 3.3). Figure 3.3 is reproduced in Hastings et al. [40] and Hastings [41] (subsequent) with theoretical predictions superimposed. The ground and flight data reported therein are from Ferguson [39]. He concluded the following (and we further comment in parenthesis):

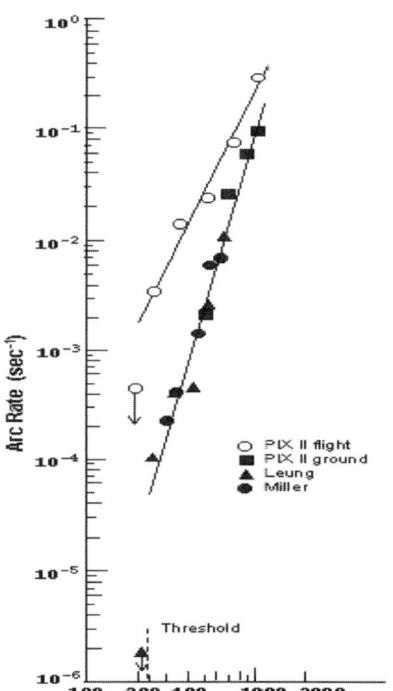

1. A threshold for arcing of 2 × 2-cm solar cells into the plasma appears to exist near -230 V (with respect to the plasma). A threshold can exist for 5.9 × 5.9-cm cells at a lower voltage, but is not yet proven. (More modern studies have found thresholds as low as 75 V for specific array designs. The difference in threshold is more likely due to cover-slide thickness than cell size.)

2. The arc rate at voltages above the threshold seems to be a power law of

**FIG. 3.3 Arc rate vs voltage for standard-interconnect cells, normalized to a common ion flux. Threshold is inferred from the plasma arcing measurements.**

the voltage. This, combined with a nearly linear dependence of arc rate on plasma density, produces an apparent "threshold" that varies with plasma density. (Here, "above" means for voltages more negative than the threshold voltage. The apparent threshold is just because the "waiting time" for an arc to occur has exceeded the measurement interval.)

3.  The arc rate decreases to a steady value on a timescale of a few hours. It is not yet clear whether this is due to repeated arcing or to exposure to the plasma. (Further studies have shown [7, 42] that this is due to both causes—outgassing into the vacuum removes contaminants over time, and arcs destroy contaminant islands in their burst of plasma.)

4.  The arc rate might depend on the plasma density to the first power, on the square root of the ion temperature, and inversely on the square root of the ion mass (that is, on the ion flux onto the sample).

5.  No significant dependence of the arc rate on the number of cells or interconnects could be found in the data. (This is still the case: the most likely arc site goes first, but there is no dearth of other arc sites when the charge builds back up. That this occurred in the data showed that each arc nearly completely discharged the available capacitance. Schemes can be proposed to prevent an arc from communicating with other cells or strings than the one on which it occurs, but in general all electrically connected cells or strings will contribute capacitance-stored energy to the discharge.)

6.  The arc rate is greater in the flight-test conditions than in ground tests, possibly because of the atomic-oxygen plasma in LEO. (It is unclear what other differences affect the arc rate, although cell temperature is clearly important in subsequent flight data, such as PASP-Plus.)

7.  The arc rate in cells with exposed conductors on the backs, as in welded-through substrates, is higher at all likely arcing voltages than the rate for cells exposed to the plasma only on the fronts. (This might be a result of the copper exposed on the backs, as contrasted with silver on the fronts).

Studies by Upschulte et al. [43] and Hastings et al. [44] confirm that a voltage threshold exists for solar-array arcing, and for certain values of a parameter called the field enhancement factor (FEF; see Cho et al. [45]), reasonable values of the threshold are predicted. Vayner et al. [28] (and subsequent) have shown that arcing is enhanced primarily by the presence of desorbing contaminant layers, although thin cover slides and other geometrical factors can also enhance the electric field and lower the arc threshold. Snyder et al. (Snyder, D. B., Vayner, B. V., and Ferguson, D. C., private communication, 1998) have shown that hot arrays (100°C) have a higher arc threshold than cool arrays (room temperatures) in ground tests, presumably because the cover slides become more conductive at high temperatures. These results were confirmed on orbit in the PASP Plus experiment for the APSA-type solar arrays [46].

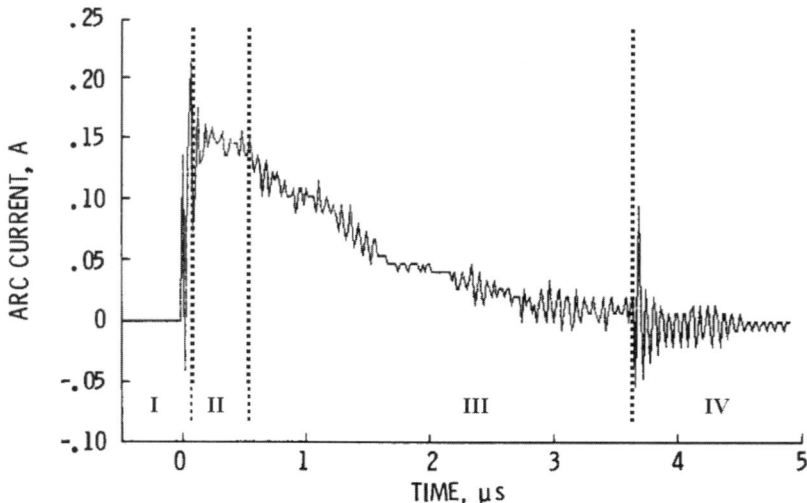

**FIG. 3.4   Typical waveform for an arc.**

### 3.3.2.1.4   Typical Waveform

Figure 3.4 [31] shows the time dependence of the current from an array segment during an arc. A typical arcing sequence has four regions:

I. The arc is initiated, and the current increases to a peak value. The rise time varies from less than 0.1 $\mu$s to about 1 $\mu$s. The peak amplitude and rise time depend primarily on the capacitance electrically connected to the arc site.

II. The current then remains near the peak value for some time.

III. The current decreases with a roughly exponential decay. The decay time associated with the termination of the arc should not be confused with the total duration of the arc. During this decay, the current is space charge limited.

IV. The arc terminates suddenly, and the array begins to recharge to the bias voltage. At this point the cover slides of the array are substantially positive relative to both space and the arc point. The cover slides collect a substantial electron current from the plasma, resulting in the observation of a slight negative pulse.

### 3.3.2.1.5   System Response

Arc currents can flow out into the surrounding plasma, with the return currents distributed over wide areas of other spacecraft surfaces.

During an arc, two things will happen. As charge leaves during an arc, the potential of the arc site changes, and the potential of the system, electrically connected to the arc site, will change. As a result of the potential change, return currents will flow to restore equilibrium. The return currents will come both from the surrounding plasma and from the arc-generated plasma. There are two impacts on other systems. The structure currents will look like noise to instrumentation. And, the change in spacecraft ground will affect plasma currents to surfaces. In principle, these responses are the same for transients of any cause: docking, thruster firings, waste dumps, and beam experiments. Only the magnitudes will be different.

The response of a system to an arc can be estimated from a circuit analysis including terms to approximate the capacitances of the surfaces to space. An arc can be simulated in such a model by injecting an appropriate current pulse and computing the circuit transients [47].

### 3.3.2.1.6 Damage potential

Initial indications that sustained arcs could cause substantial damage to solar arrays were obtained in testing where the bias power supply, intended to impress a potential difference between an array and its cover slides, was not sufficiently isolated from the sample when arcs occurred (see Continuous Arcs, Sec. 3.3.2.2.2). Tests at LeRC (now GRC) in the 1980s [48] showed that solar-array interconnects could be melted by arc currents as large as 40 A.

Although pictures of damage produced by on-orbit sustained arcs are rare, because most arrays that have arced are not recovered, we do have photos of damage suffered by the ESA Eureca spacecraft that was recovered by the space shuttle. Figure 3.5 shows a sustained-arc site on its solar arrays. In this case, the sustained arc eventually burned through the array substrate to the grounded backing, completely shorting the array string to ground.

The Space Systems (SS) Loral satellites PAS-6 and Tempo-2 underwent sustained arcing in GEO that led to several shorted solar-array strings and a severe loss of power. Although these were GEO failures, after the initial arc occurs it is believed that the mechanism for sustained arcing is the same for LEO. Subsequent SS/Loral satellites underwent extensive modification to prevent sustained arcing and have had no similar string failures since that time.

A sustained arc on a test sample of arrays for the EOS-AM1 satellite (now known as Terra) was seen in laboratory testing. In Fig. 3.6, we see a

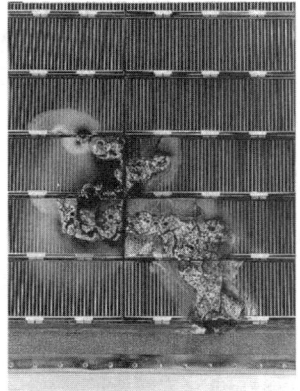

**FIG. 3.5    Sample of flight array from ESA EURECA mission after sustained arcing.**

**FIG. 3.6   Video frame from EOS-AM1 sustained-arc test.**

frame from the videotape taken during the test, and in Fig 3.7 is the vicinity of the site where the arc occurred. The capacitor used in this test to start the initial arc was 5 μF, and the arc started and continued until the power supply was manually shut off seconds later. The solar-array string was completely shorted out. This test led to rework of the entire array strings on the Terra satellite to prevent arcing on orbit.

The most famous sustained-arc event of all led to the breakage of the TSS-1R electrodynamic tether and the loss of the attached satellite. Figure 3.8 shows the burned, frayed, and broken tether end still attached to the shuttle after the break. Incidentally, the tether continued arcing long after it and its satellite were drifting free, until finally it went into night conditions where the electron density

**FIG. 3.7   Arc site of sustained arc on EOS-AM1 sample array. Cells are 2 × 4 cm.**

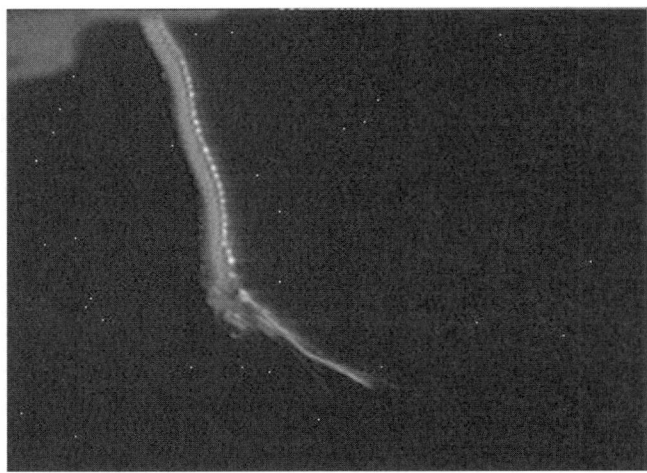

FIG. 3.8    End of the remaining TSS-1R tether.

was insufficient to sustain the arc. Noel Sargent (Sargent, N. B., private communication, 2002) has investigated whether the TSS-1R arc was seen to disrupt shuttle communications. Although he has found no record of disturbed communications during the event, for most of the time the arc was shielded by metallic structures from the communications antennas, and when the tether broke, the arc was many meters from the receiving antennas. It remains to be seen whether sustained arcs produce radio noise severe enough to be a communications problem.

When the structure or array capacitance electrically connected to the arc site is sufficiently large, the initial transient arcs themselves can be large enough to produce significant damage. In Fig. 3.9, we see an anodized aluminum plate that has undergone repeated arcing in the laboratory with the ISS structure capacitance attached. Its thermal properties have been completely destroyed, along with most of the insulating surface layer of aluminum oxide.

### 3.3.2.2    EMI

Solar-array arcs typically involve the violent discharge of very large currents for very short times.

FIG. 3.9    Anodized aluminum plate after repeated arcing [49].

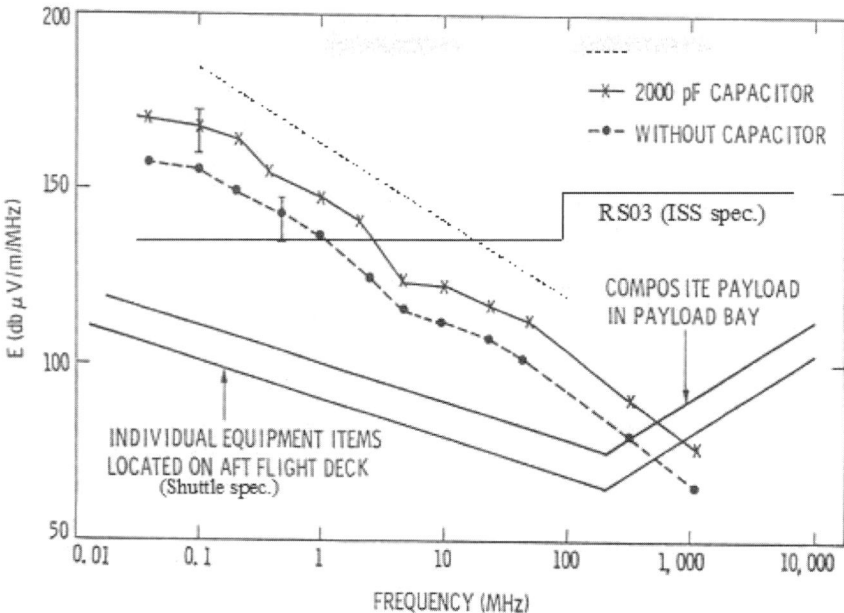

**FIG. 3.10   EMI from a small solar-array arc and a hypothetical ISS anodized aluminum arc compared to Space Shuttle Orbiter's specs.**

Not surprisingly, the electromagnetic spectrum associated with such discharges obeys the typical power law that has long been observed with arc discharges. We show an example of such a spectrum in Fig. 3.10 [50]. The test article was a small solar-array sample that was proposed for a plasma interactions experiment in the space shuttle cargo bay. The test was designed to learn whether the radiated EMI from the sample would exceed orbiter specifications. The test was done with the bare array alone and with an added capacitance that simulated the energy storage associated with a full-size array. The biasing power supplies were electrically isolated from the arcs by a large resistor. As the curves show, even arcs from a small test array exceed allowed EMI specifications over most of the frequency range. It should be expected that arcing will always produce detectable EMI and that laboratory testing will be needed to quantify the level of interference. The magnitude of radiated EMI is a strong function of the "antenna gain" composed of those conductive (radiating) elements connected to the arc site. This heavily influences the shape of the radiated EMI spectrum (Sargent, N. B., private communication, 2002). Because antenna gain is extremely difficult to estimate, testing is essential.

### 3.3.2.2.1 Structure Arcing

Generally speaking, there are two forms of structure arcing, triple-point arcing, as has been discussed for solar arrays, and dielectric breakdown. For triple-point arcing, insulator must surround a highly negative conductor, and an arc can occur at the conductor-insulator-plasma conjunction. Dielectric breakdown is completely different and will be discussed next.

An insulator not in the wake in LEO will come through current balance to a potential a few volts lower (or smaller) than the plasma potential. If that insulator covers a conductor, the conductor can be at a very different potential (for instance, such as the negative floating potential of the spacecraft). In this case, a thin insulator can undergo dielectric breakdown under the high electric field developed across it. Although this can occur for any type of insulator, it is of perhaps greatest interest in the case of anodized aluminum, the main ISS structural element, and a material used in astronaut EMUs (spacesuits). Because the dielectric layer in anodized aluminum is typically very thin (0.1–1 mil), it can break down at potentials of 100 V or less, less than the floating potential that is possible for a 160-V array. It was the arcing threat from the ISS anodized aluminum that forced ISS to incorporate the PCUs to control ISS floating potentials. The PCUs act by creating a large localized plasma cloud that makes good electrical contact with the surrounding plasma and essentially by brute force grounds the ISS structure to the ambient plasma. A generic plasma-contacting device is called a plasma contactor.

Different samples of anodized material break down at different potentials in a plasma (see [51]). While ISS sulfuric acid anodize withstands about 200 V before breaking down, the chromic acid anodize was found in ground tests to break down at about 72 V. Most disturbing of all, chromic-acid anodized samples for astronaut EMUs were found to break down at potentials of only −60 V, relative to the plasma, with a two-sigma error bar of 10 V. It is thus possible that an astronaut, grounded to ISS by his tether or conductive tools, could undergo an arc at only −50 V. A sneak circuit analysis showed that such arcs could put 1 A of current through an astronaut's heart. Because 0.1 A is enough to cause heart stoppage, it is imperative that if the ISS plasma contactors are inoperable during astronaut EVAs, a method could be used to prevent ISS astronaut workplaces from floating more than 50 V negative.

Dielectric breakdown currents will essentially discharge all surfaces close enough (about 3 m or so) for the induced plasma cloud to reach. For thin dielectric layers, a few square meters of surface are effectively a capacitor of many microfarads and can hold several joules of energy, all of which can be discharged in the arc. For many ISS surfaces, peak arc strengths of hundreds of amperes have been calculated. Arcs this strong will melt the arc site and spew molten metal through space. Plasma chamber tests of this kind of arcing are spectacular indeed! Arcs on one anodized surface have been seen to trigger arcs on nearby line-of-sight surfaces.

Very thin dielectric layers will have a low enough resistance that for the purposes of the plasma they would collect current rather than building it up on their surfaces. Thus, while mitigating dielectric breakdown, they must be considered as conductors rather than insulators.

Predicting arc thresholds for thin insulating layers is not as simple as using the published dielectric strengths for insulating materials. It has been found that identical thicknesses of the same anodization can differ by a factor of three or more in arc threshold voltage in a plasma. This may be because of differences in sealing the anodize surfaces, which could affect their resistance to plasma currents. Until the theoretical situation is better understood, plasma testing must be used to determine the dielectric strength of insulators in applications, which could lead to charging in LEO (see [51]).

Carruth et al. [52] have found that dielectric breakdown can also be initiated by simulated micrometeoroid strikes at voltages as low as 55 V.

### 3.3.2.2.2  Continuous Arc (Sustained Arc)

Arcs that occur in air when electrical contacts are made or broken are due to breakdown of the neutral gas. Although these can become continuous ("showering arcs"), they are not the same phenomenon as the continuous arcs in a LEO environment, which involve breakdown of the gas liberated by the arc itself. (See [53] for a discussion of continuous arcs in air.)

When the LEO arc circuit includes the solar arrays, distribution cabling, or other source of power, it might be possible for structure or solar-array arcs to become continuous (or sustained). Such continuous arcs, fed by the power supply, have an essentially inexhaustible source of energy and can lead to catastrophic damage. This hypothesis for the loss of solar-array strings on the SS/Loral satellites PAS-6 and Tempo II was confirmed by ground tests done by Snyder et al. [54]. Later testing on the EOS-AM1 arrays showed that continuous solar-array arcs could occur in a LEO environment at a string voltage as low as 100–120 V. (In those tests, the sustained arc occurred at a voltage relative to the surrounding plasma of −250 V.) The most recent data [55, 56] have shown that strings with potentials as low as 40 V with respect to each other can lead to sustained arcing. The scenario for the catastrophic loss is given in Ferguson et al. [57] and is summarized here.

First of all, an ordinary solar-array arc must get started. This will usually be at a triple point as just described. In the case of the SS/Loral arrays, the differential voltage between solar array and plasma could have been as low as 100 V, for the SS/Loral arrays were using thin cover slides similar to the APSA cells, which arced at voltages as low as 75 V on orbit (PASP Plus results [46]).

When the initial arc (sometimes called the trigger arc) is generated, it discharges only the local capacitance, but the arc plasma expands out from the arc site and comes in contact with an exposed conductor at a very different voltage. In the case of the SS/Loral arrays, the most positive end of the array strings was less than a millimeter away from the negative end. Now the arc plasma makes direct contact with the other conductor and makes for an almost dead short to that spot. The arc current has changed from one that is discharging capacitance to a current between two ends of the solar-array string.

If the current available to the arc site from the functioning arrays is greater than a certain threshold value (believed to be about 1 A) and the voltage between strings is above a certain value (believed to be about 40 V), the arc can become continuous. In ground tests these arcs continued until the source of power was artificially turned off. In space, presumably the arc would continue until the exposed conductors were melted through and the circuit was thereby interrupted. This could take seconds or minutes. Ground tests have shown that an arc that persists for more than a few hundred microseconds will not shut off by itself.

An arc that lasts long enough will locally heat the substrate and release gases. In the case of a Kapton® substrate, the Kapton® chars, but the char is also a good conductor and provides a path for the arc to continue. Snyder et al. [54] have shown that the heat generated in continuous arcs on Kapton® is sufficient to produce the Kapton® charring measured after the event.

In any event, a continuous arc can destroy a whole string (if the arc is between traces on the same string) or adjacent strings (if the arc is between strings) or the entire array power (if the arc is between combined power traces). The possibility of losing the entire array power on the Deep Space 1 mission caused the builders to remove a solar panel that had already been installed to modify it and its sister array to prevent continuous arcing. Its power traces were only a few millimeters apart and were exposed both to the plasma and to each other before the modifications were made. Afterwards, insulating material prevented arc plasma from shorting out between the power traces.

Anodized aluminum structure elements might be subject to continuous arcing if the arc plasma generated can contact the solar array or other power source or if the potential at the arc site can be maintained at a high enough negative level by a high-voltage electron-collecting power source. Such continuous anodized breakdowns were called "sizzle arcs" by the team that discovered them [58].

Finally, an arc on an electrodynamic tether can become continuous. The infamous arc on the TSS-1R tether that led to its break and the loss of the satellite was a continuous (sustained) arc with its power supplied by the tether. The arc site was a flaw in the tether insulation that spewed gas out which became ionized and completed the arc circuit path [59, 60]. Because in this case the power source was more one of constant voltage, rather than constant current, the 3500-V potential difference between the tether top and bottom caused the arc site to float at just the negative potential (about $-600$ V) necessary to keep the arc going and still collect the $\sim 1$-A arc current of electrons on the satellite. Had TSS-1R used a tether of greater resistance, the threshold arc current could not have been maintained. For example, a total tether resistance of 10,000 $\Omega$ would have limited the arc current to less than 0.4 A, less than the sustained arc threshold. Alternatively, if the satellite electron collection capability had been limited to less than an ampere, the arc could not have been sustained. Of course, these measures would have severely restricted the power or propulsion that could be obtained by tether operation and could not be tolerated on an experiment that was not just a proof of concept. An arc-detection circuit could have also been used to shut the tether down at the satellite end when very large currents were first detected. One should never assume that a high-voltage power system will not arc.

## 3.4  MITIGATION TECHNIQUES

### 3.4.1  CURRENT COLLECTION

If a spacecraft has no exposed high-voltage conductors, it will not collect much current. That is, insulation or encapsulation is a valid technique for preventing current collection in LEO. The GEO Spacecraft Charging Guidelines [61] recommend coating spacecraft surfaces with conducting materials to keep all surface potentials the same and reduce differential charging. In LEO, however, the space plasma will act to keep insulating surfaces at the same potential (discounting wake effects), and so conductive coatings are not needed. If encapsulation or insulation is not possible, hiding conductive surfaces (like the edges of solar cells) from the ambient plasma by use of narrow spacing of overlying insulators (like cover slides) can choke off most current collection. It has often been remarked that if the ISS solar arrays had just a little more cover-slide overhang and/or a little tighter cell spacing there would never have been an issue with ISS charging. Of course, if all high-voltage components are inside a sealed pressure vessel, they cannot collect current from the ambient plasma.

Encapsulation (or grouting with RTV) of solar arrays has been shown to be an effective method to prevent electron collection and charging [62]. One must be careful with the use of encapsulants, however, when the possibility exists of outgassing in the presence of high-voltage components. For instance, on SAMPIE, one of the high-voltage power supplies was destroyed by a Paschen discharge that occurred on a high-voltage component where the encapsulant had delaminated and a neutral pressure was enclosed with the high-voltage component (see [63] and Fig. 3.11 for Paschen curves). On TSS-1R, the "trigger arc" was a Paschen discharge due to entrained gas inside the tether pulley casings [59, 60]. In this case, a flaw in its insulation exposed the tether conductor.

Placing plasma-current-collecting conductors into the wake of a large spacecraft is an effective technique for preventing current collection. On ISS, for instance, FPP data showed that when the arrays were turned into their own wakes, they collected such a small amount of electron current that the ISS structure would not charge. On ISS, this technique of wake-pointing the arrays is now used as a backup for the plasma contacting units during astronaut EVAs.

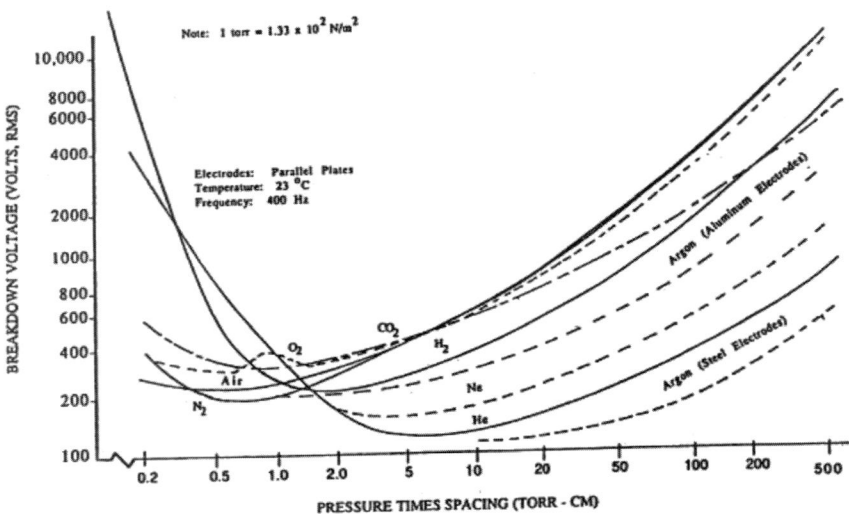

**FIG. 3.11   Voltage breakdown of pure gases as a function of pressure times spacing.**

Of course, very high potentials on wake-pointing conductors can collapse the wake, but this will require thousands of volts potential for large structures.

For a spacecraft that will often undergo auroral passage, one must be careful with the use of insulators. Like in GEO, spacecraft in the aurorae can undergo rapid differential charging on insulators, and this can lead to buildup of potentials that might lead to arcing. It is hoped that a subsequent document will cover polar-orbiting spacecraft in the detail. This chapter covers equatorial LEO orbits.

### 3.4.2   CONTROLLING SPACECRAFT POTENTIAL

There are three basic techniques to control spacecraft potential. One is to place the structure at the most positive potential generated by the LEO spacecraft power system (the positive ground option) The second is to ground the structure by brute force to the ambient plasma (the plasma contactor solution). Third, one can prevent any plasma exposure of high-voltage conducting surfaces (the encapsulation solution). We will discuss these mitigation strategies in order next. For ideas about other ways to prevent spacecraft charging, see Ferguson [64].

#### 3.4.2.1   POSITIVE GROUND

Because charging in LEO is dominated by current collection on the most positive end of the solar arrays, and the negative end floats at about 90% (typically) of the string voltage, the positive end of the array will be at about 10% of the array string voltage away from the plasma potential. For a 160-V array, this means a positively grounded structure will float at 16 V or less away from the plasma potential. Most deleterious plasma effects are minimal at such a potential. In fact, the structure in this case contributes to electron collection and will actually float closer to plasma potential than the positive end of the array will, taken alone, because of exposed grounded conductors on the structure.

However, most spacecraft power systems are negatively grounded because of a dearth of space-qualified electronics with the positive ground polarity. Although very efficient PMAD systems now exist that use buck-boost converters to change the ground polarity and voltage [65], most spacecraft busses do not incorporate this technology yet.

For instance, when ISS charging possibilities were first being considered, it was estimated that to change the power system ground from negative to positive would cost at least $100 million. It was decided instead to use the plasma contactor mitigation strategy detailed next, which ended up costing less than $35 million.

A variant of this technique uses a center-tapped array, but will only cut the maximum structure potential to about half of the solar-array string voltage.

### 3.4.2.2  PLASMA CONTACTORS

A device that makes good contact with the surrounding plasma can effectively ground its point of contact. If the device is a large sheet of metal, it will dominate current collection and stay near plasma potential. However, the sheet of conductor must be much larger than the solar-array effective-electron collecting area for this solution to work. In the case of ISS for instance, the metal sheet would need hundreds or thousands of square meters of ram-ion collecting area to be effective. In LEO, the drag produced by such a large area would be prohibitive.

Electron guns were used on PIX-II [66] and PASP Plus [67] to emit the electrons being collected by high-voltage solar arrays and thus prevent charging, but such devices are limited by space charge considerations to low emitted electron currents. A better solution is a device that is not limited by space charge considerations, that is, a plasma contactor.

A plasma contactor generates a high-density plasma cloud, which expands and makes good electrical contact with the ambient plasma. Usually a hollow cathode device [68] is used to emit a xenon plasma, whose space charge is nullified by nearly equal densities of electrons and ions in the emitted cloud. The very mobile electrons carry current into the surrounding ambient plasma. This current can be very large. For instance, the ISS PCU device has a hollow cathode element smaller than a little finger, but can emit up to 10 A of continuous electron current. In the case of ISS, the PCU acts like a ground rod at its location to effectively ground (to within about 20 V) the structure to the ambient plasma. Of course, at other points, the structure will still have the $v \times B \cdot l$ potential away from the ambient plasma. This is only 40 V from end to end on the largest structure ever orbited (i.e., ISS), and so at all points the potential is outside the arcing range ($-50$ V or less).

Although a hollow cathode plasma contactor requires xenon gas vessels, refurbishment, etc., other devices with little or no expellant are being explored for use as plasma contactors. As an example, a plasma contactor made of microtips and microscopic holes, with an imposed bias, could theoretically emit electrons over a wide area and thus defeat the space charge limitation with no working gas. A patent has been awarded for using such a device to control spacecraft potentials in GEO [69], but making such a device work reliably in the high-density plasma of LEO is no small feat and has not yet been done.

### 3.4.2.3   ENCAPSULATION

Encapsulating the high-voltage conductors on solar arrays can have a two-fold beneficial effect. First, it can prevent arcing at triple points by keeping the plasma away from the conductor-insulator junctions. Second, it can prevent electron collection by the arrays and thus prevent spacecraft charging at its root cause. So far, the only arrays ground tested in a simulated LEO plasma to withstand bias voltages greater than 300 V have been those with the arrays or cells encapsulated [38, 62, 70].

When encapsulating arrays or cells, several caveats must not be ignored. First, no air must be entrained anywhere. Although this might seem obvious, at least one set of encapsulated test arrays sent to NASA's Glenn Research Center had sufficient air entrained that the coating delaminated and swelled under vacuum. In fact, so much air was entrained that the test articles under vacuum appeared to swell up like plastic balloons. In cases where only a very small amount of air is trapped, visible effects might not occur, yet the trapped air will present the danger of Paschen breakdown under high voltage (see Fig. 3.11).

Second, the encapsulant thickness must be sufficient to withstand dielectric breakdown at the highest array voltage. For thin-film arrays, this consideration can contribute significantly to the array mass. In keeping with the discussion we had under structure arcing, it is important that thin-film encapsulants be tested under voltage in a plasma environment, rather than relying solely on published dielectric strengths.

Third, the encapsulant must not be able to peel away from high-voltage components, or Paschen breakdown can occur as a result of entrained outgassing products that can reach sufficiently high neutral pressures. Figure 3.11 [71] shows the Paschen breakdown curve for a number of gases for dc to low-frequency ac ($\sim$400 Hz) for parallel plates. Here it can be seen that for a wide range of pressure distance combinations, the Paschen minima are typically around a few hundred volts for common gases. Helium gas has the lowest Paschen minimum voltage. Most outgassing products have not had their Paschen curves measured. In the case of solar arrays, a coverglass that covers many cells must also make allowances for escape of outgassing products from adhesives, as well. It must be treated for all intents and purposes as a vented enclosure (treated next).

Fourth, the encapsulant must be able to withstand other aspects of the space environment for its design lifetime. Atomic oxygen, micrometeoroids and debris, and UV and x-ray exposure are some of the threats to the encapsulant. Glass stands up well to all of these environments. Some plastics do not.

The use of a sealed pressure vessel eliminates environmental interactions, and this applies to plasma interactions as well. In the more general case, high-voltage systems other than solar arrays are usually contained in a vented enclosure. To avoid plasma interactions, care must be taken that plasma does not enter the enclosure and react with exposed conductors inside. The key requirement on such systems is that all openings must be smaller than the plasma Debye length, which depends on the plasma density and temperature. One can readily estimate the maximum opening consistent with such a requirement.

The plasma will be capable of maintaining electric fields over a distance of approximately one Debye length $\lambda_D$, which is given by

$$\lambda_D = (kT_e/4\pi ne^2)^{1/2} = 7.43 \times 10^2 (T_e/n)^{1/2} \tag{3.4}$$

where $T_e$ is the electron temperature in electron volts, $k$ is the Boltzmann constant, $\pi = 3.14159\ldots$, $e$ is the charge of the electron, and $n$ is the electron density in negative cubic centimeters. Placing representative values from International Reference Ionosphere (IRI-86) simulations in the preceding equation, one finds a minimum Debye length from 0.12 cm at 1100 K to 0.17 cm at 2300 K.

Openings in the experiment electronics enclosure must have smaller dimensions than this minimum to prohibit plasma interactions with the experiment electronics. Larger openings can be used if covered with an electrically connected conductive wire mesh of spacing less than the minimum Debye length. To provide a reasonable margin of safety, a general guideline is that no opening should exceed 0.10 cm in its largest dimension.

## 3.4.3    ARCING

### 3.4.3.1    ON-ORBIT ARC DETECTION

Usually when ground testing solar arrays under simulated LEO plasma conditions, and especially when the array might undergo sustained arcing, an arc detection circuit is employed. It essentially looks for a rapid positive change of the array or arc site potential toward the plasma potential, as this must happen when electrons are emitted during an arc. Conversely, one can sense the emission of copious electrons and use this for arc detection. Further, the broadband EMI from an arc can be used for arc detection. In any event, electrical detection techniques can unambiguously detect an arc when it occurs. Then, in the ground tests, the power supply is electrically disconnected from the array, to prevent the occurrence of sustained arcs that might damage or destroy the sample. Sometimes, the power supply is only disconnected when the arc continues for longer than 200 ms, for example, so that arcs that would be permanently sustained can be counted, but are not allowed to wreak their damage on the sample. Such arc detection and array protection circuits can be built and used on solar arrays operating on orbit. If this is done, rather than totally preventing arcs, the damage they can do to the arc site can be limited or prevented. In this way, one can perhaps live with the arcs that do occur.

It must be obvious that the power to the LEO spacecraft will be interrupted whenever the array arc circuit is broken. Rather than being the first line of defense against arcing, arc detection and array shunting must only be used when the disruptions they cause will be infrequent.

### 3.4.3.2  PREVENTION TECHNIQUES

The design of a solar array must consider the plasma environment and interactions with that environment. Arc prevention is extremely important. The following techniques have been shown in ground and flight tests to prevent arcs or minimize their damage:

1.  If possible, use array string voltages of less than 55 V. No trigger arcs have been seen on LEO arrays of less than about 55-V string voltage even under simulated micrometeoroid bombardment. Solar arrays coming out of eclipse will generate more voltage than when they operate at their max power point.

2.  If solar-array cell edges or interconnects are exposed to the LEO plasma and string voltages are greater than 55 V, the strings should be laid out on the substrate such that no two adjacent cells have a voltage difference of greater than 40 V. Sometimes a leapfrog arrangement will be sufficient. In other high-voltage arrays, the strings should be arranged parallel to each other. Serpentine strings can be used to prevent the array width from becoming prohibitive. If the string layout cannot be modified to prevent cells with more than 40-V difference being adjacent to each other (anything less than about 1 cm can be considered adjacent), then the total string voltage must be kept low enough that the initial (trigger) arcs do not take place. The lowest known array trigger arcing has occurred on thin-coverglass cells at about 75 V ([46], see PASP Plus results).

3.  For array string voltages greater than about 75 V, trigger arcs in LEO can only be completely prevented by encapsulating the cell or array edges so that they do not see the ambient plasma. The caveats just mentioned under Sec. 3.4.1 must be followed. If encapsulation is not possible, a thorough array bakeout on orbit (1 week at 100°C or more) can get rid of contaminants and prevent trigger arcing up to about 300 V, or possibly more (see [7]). Recontamination can occur on "dirty" spacecraft (spacecraft with excessive

venting, cold-gas nozzles, etc.). Good encapsulation can prevent arcing up to a 1000-V string voltage.

4. Sustained (or continuous) arcs can occur whenever trigger arcs occur and adjacent cells have more than 40-V potential differences. However, sustained arcs, in addition to this voltage threshold, have a current threshold, below which they will not occur. It is believed that the current threshold is greater than about 1.0 A. If the current produced by each cell is above this threshold, a single string can sustain arcs. If each cell is below this current threshold, then isolating separate strings of solar cells from each other will prevent other strings from "feeding" the arc site and will prevent sustained arcs. This isolation can be achieved by using blocking diodes in each string (EOS-AM1, now called Terra, e.g., [54]). Care must be taken that the power bus and/or other components do not have the conditions necessary for sustained arcing. On the Terra arrays, for instance, it was found that diodes used to block interstring currents did not prevent the bus power traces from having sustained arcing events. Covering all exposed bus conductors with Kapton® insulation finally solved the problem. Low-outgassing RTV can be used to cover bare conductors as well.

5. RTV grout between adjacent solar cells and strings that have a high voltage with respect to each other has been shown to effectively block sustained arcs between cells and strings. The degree of coverage is important in determining the final voltage threshold for sustained arcing.

6. Arrays of 300 V and greater string voltage must be fully encapsulated in order to prevent arcing. Caveats involved under Sec. 3.4.1 must be followed.

7. Finally, although design and construction are important in preventing trigger arcs and sustained arcs, each new solar-array implementation must be tested in a simulated LEO plasma before it can be sure not to arc. This step must not be omitted. The test-bias voltage relative to the plasma should include the maximum when the arrays come out of eclipse (or the highest potential expected on the "floating" spacecraft). The interstring voltage should be at least as great as that expected anywhere on the solar array on orbit. Tests should ideally be conducted at sample temperatures as low as the eclipse-egress temperature.

8. NASA Standard 4005, "Low Earth Orbit Spacecraft Charging Design Standard," should be used when designing LEO spacecraft.

## 3.5  MODELING

### 3.5.1  SPACECRAFT CHARGING

The severity and widespread nature of plasma interactions have led to a considerable investment in the development of computer models. There are many empirical and semi-empirical models available with varying levels of capability and fidelity. Because the physics of current collection is fully embodied in Poisson's equation, a first-principles treatment is both possible and practical. The most comprehensive such code available for many years for LEO was NASCAP/LEO (NASA Charging Analyzer Program/low Earth orbit). This code was developed as a follow-on to the original NASCAP program that dealt with spacecraft charging in geosynchronous orbit.

A finite element-based solver, NASCAP/LEO reasonably approximated the geometry of sophisticated satellites or subsystems. With an expandable materials database, it iteratively solved for the potentials on all surfaces and electric fields in nearby space. The existing code was designed for mainframe and workstation deployment, made many approximations necessitated by the limited desktop computing power of the mid 1980s, and had a reputation for having a steep learning curve. It is nevertheless credited with considerable success and in the hands of a skilled user was powerful and reliable.

A new version, called NASCAP-2K, developed by NASA in conjunction with the U.S. Air Force, is now available. NASCAP-2K incorporates lessons learned over the past 25 years, takes full advantage of modern computing power (it can run on a laptop, for example) with much more sophisticated algorithms, and is designed to be easier to use. Capable of modeling current collection and charging under LEO, GEO, and auroral conditions, NASCAP-2K supersedes both NASCAP and NASCAP/LEO (see [72]).

Of special interest here is a computer-modeling tool called EWB (the Environments WorkBench, see [73]). This tool, which can run on a desktop or laptop PC, uses simple models of plasma environments and interactions to predict LEO spacecraft floating potentials, for example. Over 100 models of the LEO environment are included in this integrated code and over 50 interactions models, including the plasma interactions models considered here. EWB was extensively funded by the ISS and is the official ISS plasma interactions tool. Detailed and extensive models of various ISS configurations are included with EWB, although the code can also be used to create and model a wide variety of different LEO spacecraft. Both EWB and NASCAP-2K are subject to ITAR restrictions and at present cannot be given to non-U.S. citizens. For more information on distribution of these codes, please see http://see.msfc.nasa.gov. Similar codes produced by ESA and JAXA exist and can be found on the worldwide web.

In Fig. 3.12 is a plot showing the result of an EWB calculation of potentials on the ISS mission build 12A. Here, a special model of ISS solar-array current collection and ISS solar-array mast wire current collection, based on PCU measurements of previous ISS mission builds, was constructed by Science Applications International Corporation (SAIC). The potentials shown were determined by iteration until the current balance equation was satisfied for ISS as a whole. In this figure, the PCUs were turned off to investigate charging under PCU failure conditions. It is clear that for this configuration most of the vehicle charging is caused by $v \times B \cdot l$ effects across the long truss and solar-array segments. Not shown are

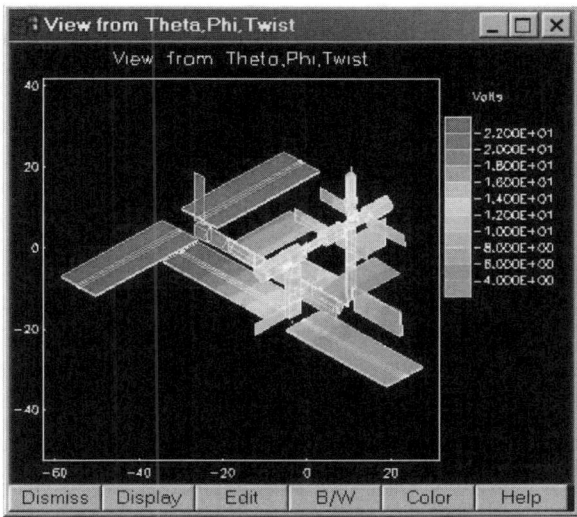

**FIG. 3.12    EWB contour plot of ISS potentials. This is the 12A mission configuration at an arbitrary point in the orbit.**

the EWB screens that detail the potentials and currents on each ISS component. EWB can also easily calculate the time dependence of all of the ISS potentials during an orbit and their dependencies on plasma parameters and changes in the detailed ISS configuration. Of course, EWB can also be used for other spacecraft. Figure 3.12 just illustrates how a very complex system can be analyzed with this extremely useful computer code.

## 3.5.2  ARCING

The process of electrical breakdown has not lent itself well to modeling, and solar arrays are no exception. The just-mentioned computer codes for determining potentials on all surfaces and electric fields in nearby space are certainly useful for solar arrays, but the actual initiation of an arc is extremely difficult to predict. Despite NASA and the Air Force's efforts to fund theoretical work in this area during the 1990s, no reliable model for arc initiation exists. Experience has shown that that knowledge of the potential distribution is at best a rough indicator of the probability of an arc. The complex geometries involved in cell construction and string layout along with the poorly understood properties of adhesives, coatings, and other materials often result in laboratory tests behaving in unexpected ways. This emphasizes the need for testing of solar arrays in suitable space environmental chambers and ultimately as part of space experiments.

## 3.6   TESTING

The importance of testing in mitigating LEO spacecraft charging and its effects cannot be overestimated (see [74]). A valid LEO arc test must take place in a vacuum of pressure less than 10 $\mu$ torr. It must generate a plasma with an electron density of more than $10^5$ electrons per cubic centimeter. The electron temperature must be less than about 3 eV, and the plasma should not be a streaming plasma. (It should be essentially isotropic.) The sample temperature must be as low as the lowest sunlit temperature on orbit. To ensure that arcs will not occur in space, a sufficiently long waiting time must be used at each bias voltage that the arc rate is measured to be statistically significantly lower than the threshold arc rate. If the threshold is unknown, see Ferguson [39] for a proper technique for establishing it in ground tests. Be aware that the arc rate at a given voltage usually decreases with time in the plasma—do not confuse this with an increasing arc voltage threshold (see [39]). The chamber used for the tests should be big enough that the plasma sheath of the biased sample does not reach the chamber walls. Finally, use solar-array design and building techniques that have been space qualified, whenever possible.

In LEO plasma testing, the array or anodized aluminum potential relative to the plasma (which in space is due to spacecraft charging) is usually obtained by biasing the sample with a dc power supply. If one is interested in investigating transient arcs, one must decouple the dc power supply from the arc current during an arc. This means the bias supply circuit must have a time constant greater than a few hundred microseconds, so that the arc can build up and dissipate without being powered by the bias supply. This can be done by putting a large resistance in the arc circuit and incorporating a capacitor to simulate the array or structure capacitance that would be discharged in the arc. For instance, if the on-orbit capacitance connected to the arc site is expected to be 0.1 $\mu$F, then this value capacitor can be used to provide current during the arc. With such a capacitor, the bias supply circuit can be given a 1-ms RC time constant (much greater than the arc timescale) by using a 10-k$\Omega$ series resistance. This effectively decouples the bias power supply from the arc. Of course, it also puts an upper limit on the arc rate attainable because of recharge time considerations.

If doing nondestructive sustained-arc testing, the series resistance should be adjusted to limit the maximum current to that expected in the arc, and a cutoff circuit should be employed to shut off the bias supply after a few hundred microseconds. Experience shows that an arc that continues under such circumstances for more than about 200 $\mu$s will be sustained. Arc current and/or voltage waveforms should be closely monitored to distinguish between transient and sustained arcs. Videotapes of arc locations are helpful for diagnostic purposes. If destructive sustained arcs are allowed to occur, the videotape can confirm the arc time duration. A new ISO standard, ISO 11221, "Space Systems – Space Solar Panels – Spacecraft Charging Induced Electrostatic Discharge Test Methods," has been agreed upon by the international community and should be used whenever possible for LEO spacecraft arcing tests.

# REFERENCES

[1]   Hedin, A., "MSIS-86 Thermospheric Model," *Journal of Geophysical Research*, Vol. 92, 1987, pp. 4649–4660.

[2]   Prag, A. B., "A Comparison of the MSIS and Jacchia-70 Models with Measured Atmospheric Density Data in the 120 to 200 km Altitude Range," NASA TR-0083 (3940-04)-1, 1983.

[3]   "Computational Procedure Used in the Development of the MSFC Modified Jacchia Model Atmosphere," *CE Environment Criteria Guidelines for Use in Space Vehicle Development*, NASA Marshall Space Flight Center, N70-40876 23-30, Huntsville, AL, 1970.

[4]   Hickey, M., "The NASA Marshall Engineering Thermosphere Model," NASA-CR-179359, 1988.

[5]   "U.S. Standard Atmosphere," NOAA, NOAA-S/T-76-1562, Washington, D.C., 1976.

[6]   King, R. L., "A Computer Version of the US Standard Atmosphere," NASA-CR-150778, 1978.

[7]   Vayner, B., Galofaro, J., Ferguson, D., and Degroot, W., "Electrostatic Discharge Inception on a High-Voltage Solar Array," NASA/TM-2002-211329, 2002.

[8]   Chen, F. F., "Electric Probes," *Plasma Diagnostic Techniques*, edited by R. J. Huddlestone and S. L. Leonard, Academic International Press, New York, 1965, pp. 113–200.

[9]   Cohen, H. A., Cooke, D. L., Evans, R. W., Hastings, D., Jongeward, G., Laframboise, J. G., Mahaffey, D., Mcintyre, B., Pfizer, K. A., and Purvis, C., "Working Group Report on Advanced High-Voltage High-Power and Energy-Storage Space Systems." *Space Technology Plasma Issues in 2001*, Jet Propulsion Lab., Pasadena, CA, 1 Oct. 1986.

[10]  Zhang, T. X., Hwang, K. S., Wu, S. T., Stone, N. H., Chang, C. L., Drobot, A., Wright, K. H., Jr., and Rose, M. F., "Satellite Motion Effects on Current Collection in Low Earth Orbit," NASA Marshall Space Flight Center, doc id: 20000110580, 1 Jan. 2000.

[11]  Stone, N. H., and Raitt, W. J., "The TSS-1R Electrodynamic Tether Experiment: Scientific and Technological Results," NASA Technical Report, NASA Marshall Space Center, Huntsville, AL, 1 Jan. 1998.

[12]  Stone, N. H., Wright, K. H., Winningham, J. D., Papadapolous, K., Zhang, T. X., Hwang, K. S., Wu, S. T., and Samir, U., "A Review of Scientific and Technological Results from the TSS-1R Mission," Tether Technology Interchange Meeting, Paper, 1 Jan. 1998.

[13]  Vaughn, J. A., et al., "Review of ProSEDS Electrodynamic Tether Development," AIAA Paper 2004-3507, July 2004.

[14]  Cooke, D. L., Talbot, J., and Shaw, G., "Pre-Flight POLAR Code Predictions for the CHAWS Space Flight Experiment," Phillips Lab., PL-TR-94-2056, Hanscom AFB, MA, 31 Jan. 1994.

[15]  Bonito, N. A., Bounar, K. H., McNeil, W. J., Roth, C. J., Tautz, M. F., and Vancour, R. P., "Spacecraft Interactions Modeling and Post-Mission Data Analysis," Radex, Inc., Rept. RXR-96081, Bedford, MA, 15 Aug. 1996.

[16]  Raitt, W. J., Siskind, D. E., Banks, P. M., and Williamson, R. R., "Measurements of the Thermal Plasma Environment of the Space Shuttle," *Planetary and Space Science*. Vol. 32, 1 April 1984, pp. 457–467.

[17]  Murphy, G., Pickett, J., Dangelo, N., and Kurth, W. S., "Measurements of Plasma
      Parameters in the Vicinity of the Space Shuttle," *Planetary and Space Science.* Vol.
      34, 1 Oct. 1986, pp. 993–1004.

[18]  Chock, R. C., "NASCAP/LEO Simulations, SSF Solar Cell Geometries," Minutes of
      the Electrical Grounding Tiger Team Meeting, Boeing Trade Zone, May 1991.

[19]  Chock, R., "NASCAP/LEO Simulations of Shuttle Orbiter Charging During the
      SAMPIE Experiment," *Fifth Annual Workshop on Space Operations, Applications,
      and Research (SOAR '91)*, CP 3127, NASA, 1991, p. 655.

[20]  Ferguson, D. C., and Gardner, B., "Modeling International Space Station (ISS)
      Floating Potentials," AIAA Paper 2002-0933, Jan. 2002.

[21]  Hillard, G. B., "Plasma Chamber Testing of Advanced Photovoltaic Solar Array
      Coupons," *Journal of Spacecraft and Rockets*, Vol. 31, No. 3, 1994, pp. 530–532.

[22]  Kennerud, K. L., "High Voltage Solar Array Experiments," NASA-CR-121280,
      March 1974.

[23]  Ferguson, D. C., Hillard, G. B., Snyder, D. B., and Grier, N. T., "The Inception of
      Snapover on Solar Arrays: A Visualization Technique," AIAA Paper 98-1045, Jan.
      1998.

[24]  Vayner, B., Galofaro, J., Ferguson, D., deGroot, W., Thomson, C., Dennison, J. R.,
      and Davies, R., "The Conductor-Dielectric Junctions in a Low Density Plasma,"
      NASA/TM-1999-209408, Nov. 1999.

[25]  Button, R., Brush, A., and Sundberg, R., "Development and Testing of a 20 kHz
      Component Test Bed," *IECEC-89. Proceedings of the Twenty-fourth Intersociety
      Energy Conversion Engineering Conference*, Vol. 1 (A90-38029 16-20), Inst. of
      Electrical and Electronics Engineers, New York, 1989, pp. 605–610.

[26]  Ferguson, D. C., "Ram-Wake Effects on Plasma Current Collection of the PIX 2
      Langmuir Probe," *Spacecraft Environment Interactions Technology*, 1985, pp. 349–
      357.

[27]  Samir, U., Stone, N. H., and Wright, K. H., "On Plasma Disturbances Caused by the
      Motion of the Space Shuttle and Small Satellites–A Comparison of in situ
      Observations," *Journal of Geophysical Research*, Vol. 91, Jan. 1986, pp. 277–285.

[28]  Vayner, B., Galofaro, J., and Ferguson, D., "Arc Inception Mechanism on a Solar
      Array Immersed in a Low-Density Plasma," NASA/TM-2001-211070, July 2001.

[29]  Purvis, C. K., Ferguson, D. C., Snyder, D. B., Grier, N. T., Staskus, J. V., and Roche,
      J. C., "Environmental Interactions for Space Station and Solar Array Design–
      Preliminary," NASA Lewis Research Center, 1986.

[30]  Jongeward, G. A., Katz, I., Mandell, M. J., and Parks, D. E., "The Role of
      Unneutralized Surface Ions in Negative Potential Arcing," *IEEE Transactions on
      Nuclear Science.* Vol. NS-32, Dec. 1985, pp. 4087–4089.

[31]  Snyder, D. B., and Tyree, E., "The Effect of Plasma on Solar Cell Array Arc
      Characteristics," NASA-TM-86887, Jan. 1984.

[32]  Snyder, D. B., "Characteristics of Arc Currents on a Negatively Biased Solar Cell
      Array in a Plasma," NASA-TM-83728, Jan. 1984.

[33]  Parks, D. E., Jongeward, G. A., Katz, I., and Davis, V. A., "Threshold Determining
      Mechanisms for Discharges in High Voltage Solar Arrays,." AIAA Paper 86-0364,
      Jan. 1986.

[34]  Galofaro, J. T., Doreswamy, C. V., Vayner, B. V., Snyder, D. B., and Ferguson, D. C.,
      "Electrical Breakdown of Anodized Structures in a Low Earth Orbital
      Environmental," NASA/TM-1999-209044, April 1999.

[35]    Malter, L., "Anomalous Secondary Electron Emission A New Phenomenon," *Physical Review*, Vol. 49, 1936, p. 478.

[36]    Cho, M., and Hastings, D. E., "Dielectric Charging Processes and Arcing Rates of High Voltage Solar Array," AIAA Paper 91-0605, Jan. 1991.

[37]    de la Cruz, C. P., Hastings, D. E., Ferguson, D. C., and Hillard, G. B., "Data Analysis and Model Comparison for Solar Array Module Plasma Interactions Experiment," *Journal of Spacecraft and Rockets*, Vol. 33, No. 3, 1996, pp. 438–446.

[38]    Brandhorst, H., and Best, S., "Hypervelocity Impact Studies on Solar Cell Modules," Auburn Univ., Rept. AU-4-21839, Auburn, AL, 25 March 2001.

[39]    Ferguson, D. C., "The Voltage Threshold for Arcing for Solar Cells in LEO: Flight and Ground Test Results," NASA TM 87259, March 1986.

[40]    Hastings, D. E., Cho, M., and Kuninaka, H., "The Arcing Rate for a High Voltage Solar Array – Theory, Experiment and Predictions," *Journal of Spacecraft and Rockets*. Vol. 29, No. 4, 1992, pp. 538–554.

[41]    Hastings, D. E., "A Review of Plasma Interactions with Spacecraft in Low Earth Orbit," *Journal of Geophysical Research*, Vol. 100, No. A8, 1995, pp. 14,457–14,483.

[42]    Galofaro, J., Vayner, B., Degroot, W., and Ferguson, D., "The Role of Water Vapor and Dissociative Recombination Processes in Solar Array Arc Initiation," NASA TM 2002-211328, March 2002.

[43]    Upschulte, B. L., Marinelli, W. J., Carleton, K. L., Weyl, G., Aifer, E., and Hastings, D. E., "Arcing of Negatively Biased Solar Cells in a Plasma Environment," *Journal of Spacecraft and Rockets*, Vol. 31, No. 3, 1994, pp. 493–501.

[44]    Hastings, D. E., Cho, M., and Kuninaka, H., "The Arcing Rate for a High Voltage Solar Array – Theory, Experiment and Predictions," AIAA Paper 92-0576, Jan. 1992.

[45]    Cho, M., and Hastings, D. E., "Dielectric Charging Processes and Arcing Rates of High Voltage Solar Array," AIAA Paper 91-0605, Jan. 1991.

[46]    Soldi, J. D., and Hastings, D. E., "Arc Rate Simulation and Flight Data Analysis for the PASP Plus Experiment," MIT, Rept. AD-A301837, PL-TR-95-2126, Cambridge, MA, Sept. 1995.

[47]    Metz, R. W., "Circuit Transients due to Arcs on a High Voltage Solar Array," *Journal of Spacecraft and Rockets*, Vol. 23, No. 5, 1986, pp. 499–504.

[48]    Miller, W. L., "An Investigation of Arc Discharging on Negatively Biased Dielectric Conductor Samples in a Plasma," *Spacecraft Environmental Interactions Technology*, March 1985, pp. 367–377; also N85-22470 13-18.

[49]    Schneider, T., Carruth, M. R., and Hansen, H., "ISS Plasma Interactions Arc Testing at MSFC," Powerpoint Presentation, NASA ISS TIM, March 2002.

[50]    Leung, P., "Characterization of EMI Generated by the Discharge of a VOLT Solar Array," NASA-CR-176537, Nov. 1985.

[51]    Hillard, G. B., Bailey, S. G., and Ferguson, D. C., "Anodized Aluminum as Used for Exterior Spacecraft Dielectrics," *6th Spacecraft Charging Technology Conference*, AFRL-VS-TR-20001578, U.S. Air Force Research Lab., 2000, pp. 111–113.

[52]    Carruth, M. R., Jr., Ferguson, D. C., Suggs, R., and McCollum, M., "ISS and Space Environment Interactions Without Operating Plasma Contactor," 39th Aerospace Sciences Meeting and Exhibit, AIAA Paper, Jan. 2001.

[53]    Holm, R., *Electric Contacts: Theory and Applications*, 4th ed., Springer-Verlag, Berlin 1999.

[54]    Snyder, D. B., Ferguson, D. C., Vayner, B. V., and Galofaro, J. T., "New Spacecraft-Charging Solar Array Failure Mechanism," *6th Spacecraft Charging*

*Technology Conference*, AFRL-VS-TR-20001578, U.S. Air Force Research Lab., 2000, pp. 297–301.

[55]  Hoeber, C. F., Robertson, E. A., Katz, I., Davis, V. A., and Snyder, D. B., "Solar Array Augmented Electrostatic Discharge in GEO," AIAA Paper 98-1401, Feb. 1998.

[56]  Vayner, B. V., Galofaro, J. T., and Ferguson, D. C., "Interactions of High-Voltage Solar Arrays with Their Plasma Environment: Ground Tests," *Journal of Spacecraft and Rockets*, Vol. 41, No. 6, 2004, pp. 1042–1050.

[57]  Ferguson, D. C., Snyder, D. B., Vayner, B. V., and Galofaro, J. T., "Array Arcing in Orbit - From LEO to GEO," AIAA Paper 99-0218, Jan. 1999.

[58]  Murphy, G., Croley, D., Ratliff, M., and Leung, P., "The Role of External Circuit Impedance in Dielectric Breakdown," AIAA Paper 92-0821, Jan.1992.

[59]  Szalai, K. J., Bonifazi, C., Joyce, P. M., Schwinghamer, R. J., White, R. D., Bowersox, K., Schneider, W. C., Stadler, J. H., and Whittle, D. W., "TSS-1R Mission Failure Investigation Board," NASA-TM-112426, May 1996.

[60]  Vaughn, J. A., McCollum, M. B., and Kamenetzky, R. R., "TSS-1R Failure Mode Evaluation," *Thirty-First Aerospace Mechanisms Symposium*, NASA Marshall Space Flight Center, May 1997, pp. 309–320.

[61]  Snyder, D. B., "Dynamic Interactions Between Ionospheric Plasma and Spacecraft," *Radio Science Bulletin*, No. 274, Sept. 1995, p. 29.

[62]  Reed, B. J., Harden, D. E., Ferguson, D. C., and Snyder, D. B., "Boeing's High Voltage Solar Tile Test Results," 17th Space Photovoltaic Research and Technology Conference, Sept. 2001.

[63]  Ferguson, D. C., and Hillard, G. B., "Lessons for Space Power System Design from the SAMPIE Flight Experiment," AIAA Paper 97-0087, Jan.1997.

[64]  Ferguson, D. C., "Alternatives to the ISS Plasma Contacting Units," AIAA Paper 2002-0934, Jan. 2002.

[65]  Button, R. M., Kascak, P. E., and Lebron-Velilla, R., "Digital Control Technologies for Modular DC-DC Converters," NASA/TM-2002-211369, Feb. 2002.

[66]  Purvis, C. K., "The Pix-II Experiment: An Overview," *Spacecraft Environmental Interactions Technology 1983*, U.S. Air Force Research Lab., NASA CP-2359, AFGL-TR-85-0018, 1985, pp. 321–332.

[67]  Guidice, D. A., Davis, V. A., Curtis, H. B., Ferguson, D. C., and Hastings, D. E., "Photovoltaic Array Space Power Plus Diagnostics (PASP Plus) Experiment," MIT, Rept. AD-A331959 PL-TR-97-1013, Cambridge, MA, March 1997.

[68]  Davis, V. A., Katz, I., Mandell, M. J., and Parks, D. E., "Three Dimensional Simulation of the Operation of a Hollow Cathode Electron Emitter on the Shuttle Orbiter," NASA, AIAA, and PSN International Conference on Tethers in Space, Paper 16, Sept. 1986.

[69]  Katz, I., "Spacecraft Solar Array Charging Control Device," U.S. Patent # 6,177,629, 2001.

[70]  Ferguson, D. C., Hillard, G. B., Vayner, B. V., and Galofaro, J. T., "High Voltage Space Solar Arrays," 53rd International Astronautical Congress of the International Astronautical Federation, IAC Paper 02-IAA.6.3.03, Oct. 2002.

[71]  Dunbar, W. G., "Design Guide: Designing and Building High Voltage Power Supplies," Vol. II, Air Force Wright Aeronautical Labs., AFWAL-TR-88-4143, Dayton, OH, Aug.1988.

[72]  Neergaard, L. E., Minow, J., McCollum, M., Cooke, D., Katz, I., Mandell, M., Davis, V., and Hilton, J., "Comparison of the NASCAP/GEO, POLAR, SEE Charging

Handbook, and NASCAP-2K.1 Spacecraft Charging Codes," *Proceedings of the 7th Spacecraft Charging Technology Conference*, Noordwijk, The Netherlands, April 2001.

[73]  Chock, R., and Ferguson, D. C., "Environments Workbench – An Official NASA Space Environments Tool," *Proceedings of the 32nd Intersociety Energy Conversion Engineering Conference*, Inst. of Electrical and Electronics Engineers,Washington, D.C., 1997, pp. 753 – 757; also Paper IECEC-97452.

[74]  Ferguson, D. C., "The Role of Space Plasma Simulation Chambers in Spacecraft Design and Testing," *31st Intersociety Energy Conversion Engineering Conference Proceedings*, Inst. of Electrical and Electronics Engineers, Washington, D.C., 1996, pp. 2188 – 2192.

# CHAPTER 4

# Surface Discharge on Spacecraft

Mengu Cho[*]

*Kyushu Institute of Technology, Kitakyushu, Japan*

## 4.1  INTRODUCTION

Electrostatic discharge (ESD) has been one of the major causes of satellite anomalies since the early days of spaceflight. Its importance manifested itself when the practical use of GEO accelerated in the 1970s. Research and development on high-power space systems in LEO in the 1980s raised strong concerns about arcing due to high-voltage power operation in the LEO plasma environment. In the middle of the 1990s, the increased demand for broadband and satellite TV channels led to the introduction of high-power telecommunication satellites to the market. Generally speaking, the power transmission voltage is proportional to the square root of the transmitted power, in order to suppress joule loss during transmission. As a result, 100-V-bus technologies were adapted for many commercial GEO telecommunication satellite in the middle of the 1990s. The use of high-voltage bus, however, led to a series of accidents caused by ESD in orbit, sometimes leading to a complete loss of satellite functions. Since then, research on ESD on spacecraft surface has been carried out vigorously all over the world. In this chapter, the mechanism of ESD and secondary arcing is described in detail, starting from its root cause, spacecraft charging. Internal ESD, an increasingly important subject as more spacecraft orbit through the radiation belts, is also briefly described.

## 4.2  POTENTIAL GRADIENT AT TRIPLE JUNCTION

Electrostatic discharge (ESD) on a spacecraft surface originates mostly from the edge of a conductor that is adjacent to an insulator. We often call such an edge a "triple junction," which is a point where three materials with different conductivities meet. For the present case, the surface insulator, the conductor, and the

---

[*]Director, Laboratory of Spacecraft Environment Interaction Engineering, 804–8550. Senior Member AIAA.

surrounding plasma form the triple junction. Although efforts are made to make the spacecraft surface as conductive as possible, insulator materials still occupy a large fraction of the spacecraft surface. Examples of surface insulators are solar-array coverglasses, solar paddle face-sheets (mostly made of polyimide), wire harness jackets, adhesives, thermal insulators, optical solar reflectors, and so on. Wherever the boundary with a conductor material is exposed to space, there is a risk of ESD.

The average electric field at the triple junction is given by

$$E = \frac{\phi_d - \phi_{sc}}{d} \tag{4.1}$$

where we divided the difference between the insulator potential $\phi_d$ and the spacecraft body potential $\phi_{sc}$ by the thickness of insulator $d$. Because we cannot "ground" the spacecraft circuit in space, the spacecraft body acts as a grounding point. The negative end of the power circuit, such as the solar array or battery, is usually connected to the grounding point. The conductive parts on the spacecraft exterior are usually grounded to the spacecraft body.

The spacecraft body potential $\phi_{sc}$ is determined so that the net flux of negative and positive charges balances each other. Although the spacecraft body potential with respect to the surrounding plasma is an important parameter to determine the nature of the spacecraft charging phenomena, the difference between the insulator potential $\phi_d$ and the conductor potential $\phi_{sc}$ is more important in terms of ESD. The insulator surface can have a different potential from $\phi_{sc}$ because the local balance of the charged particle fluxes determines its potential. This is called differential charging.

Differential charging on the spacecraft exterior surfaces builds up an electric charge on the insulator surfaces. The electrostatic energy stored on the insulator is partially or fully released once the electric field on any part of the surface exceeds the discharge inception threshold. The ESD current path can take three different forms. The first is *blowoff*, which is the emission of negative charges (electrons) into space. The blowoff current discharges the electric charge stored in the capacitance between the spacecraft and the ambient plasma, the *absolute capacitance*. The second is *flashover*, which is a surface discharge propagating from a starting point as the surface of the dielectric becomes conductive (by the creation of a plasma). The flashover current discharges electric charge stored in the capacitance between the surface of an insulator and the spacecraft ground, the *differential capacitance*. The third is *punch-through*, which is the classical breakdown of a dielectric material, such as the breakdown of a capacitance, for example. The punch-through discharge discharges not only the electric charge stored in the capacitance between the insulator and spacecraft ground, but also the charge stored inside the insulator material due to penetrating charged particles.

When the insulator potential is more positive than the spacecraft body potential, that is, $\phi_d - \phi_{sc} > 0$, the situation is called the inverted potential gradient (IPG). When the insulator is more negative, that is, $\phi_d - \phi_{sc} < 0$, the situation

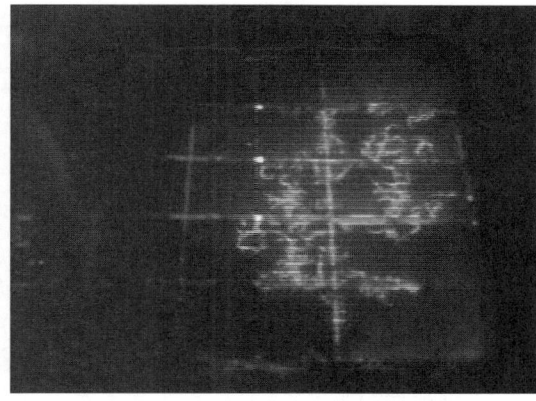

FIG. 4.1 Video image of ESD on solar array under NPG situation.

is called the normal potential gradient (NPG). ESD occurs when the potential difference exceeds a certain threshold. The inception threshold for ESD is often measured by the potential difference between the insulator surface and the spacecraft ground. Although the threshold depends upon numerous parameters such as polarity, geometry, material, ambient atmosphere, etc., generally speaking the threshold of NPG is higher than that of IPG, probably because for IPG the conductor serves as a cathode that can easily emit the electrons necessary for ESD initiation via field emission or other mechanisms.

Figures 4.1 and 4.2 show the appearance of ESD over a solar-array coupon taken by a video camera during a laboratory experiment. In Fig. 4.1 we irradiated the coupon with a 15-keV electron beam while grounding the solar-array circuit. The coverglass potential was approximately −7 kV, a NPG situation. In Fig. 4.2 we irradiated the coupon with an 8-keV electron beam while biasing the solar-array circuit to −5 kV. The coverglass potential was approximately −4 kV, an IPG situation. Surface flashover that extends over the insulator surface accompanies ESD for both the IPG and NPG cases, though its propagation characteristics differ [1].

Since the 1970s, there have been numerous ground studies of ESD under NPG situations on GEO satellites. One example is given in [2]. For IPG, most of studies were focused on LEO satellite solar arrays to prepare for high-voltage space systems such as Space Station *Freedom*. A thorough review of work done during those days is found in [3]. There were only a few studies about IPG on GEO satellites [4–7] during the 1980s. Renewed interest in IPG on GEO satellites was awakened since the end of the 1990s

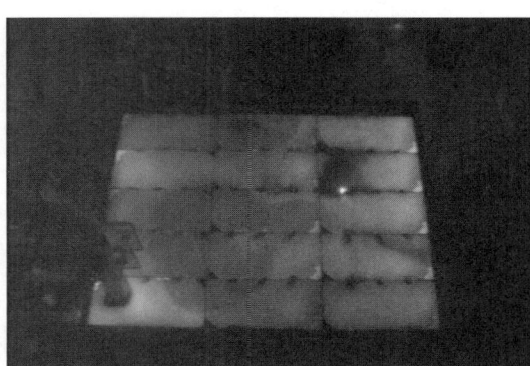

FIG. 4.2 Video image of ESD on solar array under IPG situation.

when many GEO telecommunication satellites began to fail because of ESD on their solar arrays. Since then, IPG ESD on solar arrays has been more thoroughly investigated than NPG ESD. In PEO and GEO, the secondary electron and photoelectron coefficients of insulator materials determine the polarity. As the solar array is illuminated by sunlight during power generation, and the coverglass coating such as $MgF_2$, has a very high secondary electron coefficient, IPG is expected to occur more frequently than NPG.

Whether in a normal or inverted gradient situation, ESD originates from a microdischarge at its inception point. For IPG it is believed to be as a result of the field emission and desorption of gas near the triple junction. For NPG, it might be caused by punch-through of a thin dielectric or an electron avalanche triggered by seed electrons produced by various reasons. Because of the negative potential of the spacecraft body and the high mobility of electrons, the blowoff current quickly discharges the absolute capacitance. The flashover current follows the blowoff by neutralizing the electric charge on the insulator near the ESD inception point. In the case of IPG, the flashover current flows to the inception point and forms cathode spots transforming the ESD into a kind of vacuum arc. Therefore, IPG ESD is often called a primary arc. The primary arc receives its energy from the electrostatic energy stored in the differential capacitance, which is generally much greater than the absolute spacecraft capacitance.

## 4.3   SURFACE DIFFERENTIAL CHARGING

### 4.3.1   SURFACE DIFFERENTIAL CHARGING IN GEO

In GEO, major current sources that determine the spacecraft body potential $\phi_{sc}$ are photoelectrons, ambient electrons, and secondary electrons. Under the non-substorm and noneclipse condition in GEO, outgoing photoelectrons dominate the flux of the charged particles. Therefore, the spacecraft body potential $\phi_{sc}$ becomes positive on the order of the photoelectron temperature, usually a few electron volts. Where the sun illuminates the insulator as well, its potential is also a few electron volts positive. Therefore, no serious differential charging occurs. When the insulator is in shadow, its potential can become negative comparable to the ambient electron temperature, on the order of kilovolts. Then the NPG situation appears.

When a substorm occurs in GEO, the flux of electrons of tens of kilovolts of energy increases drastically. Even during the noneclipse period, depending on satellite attitude, the ambient electron flux can exceed the photoelectron flux, and the spacecraft body potential can drop very negatively. The capacitance of the spacecraft body with respect to the surrounding plasma is of the order of 100 pF assuming that a spacecraft is a conductive body of several meters. With a current density of 10 $\mu A/m^2$ and the spacecraft surface area of 10 $m^2$, it takes only 1 ms to charge the spacecraft body to $-1$ kV. Therefore, the body potential quickly reaches the steady state within a timescale much shorter than a second.

The insulator potential needs a longer time to reach the steady state. By defining the differential voltage $\Delta V$ between the body potential $\phi_{\text{sc}}$ and the insulator potential $\phi_d$ by

$$\Delta V \equiv \phi_d - \phi_{\text{sc}} \tag{4.2}$$

we can write the temporal variation of $\Delta V$ as

$$\frac{\mathrm{d}\Delta V}{\mathrm{d}t} = \frac{1}{C_d}(j_e + j_{\text{es}} + j_{\text{ph}} + j_s) \tag{4.3}$$

where only the dominant currents are included. They are the ambient electron current $j_e$, the electron-induced electron current $j_{\text{es}}$, the photoelectron current $j_{\text{ph}}$, and the leakage current through the insulator $j_s$. These currents have a positive sign when they leave positive charge on the insulator surface. The capacitance $C_d$ is the capacitance per unit area between the insulator surface and spacecraft ground.

The currents $j_e$ and $j_{\text{es}}$ are both functions of surface potential $\phi_d$. The electron-induced secondary electron current $j_{\text{es}}$ has the following relation with $j_e$:

$$j_{\text{es}} = -\gamma_{\text{ee}}(E_e, \theta)j_e \tag{4.4}$$

where the secondary electron emission yield $\gamma_{\text{ee}}$ depends on surface material property, electron incident energy $E_e$, and incident angle $\theta$. The photoelectron current density $j_{\text{ph}}$ also depends on surface material properties and the solar incident angle. The leakage current $j_s$ is given by

$$j_s = -\frac{\phi_d - \phi_{\text{sc}}}{\rho_b d} + \frac{1}{R_s}\left(\frac{\partial^2 \phi_d}{\partial x^2} + \frac{\partial^2 \phi_d}{\partial y^2}\right) \tag{4.5}$$

where $\rho_b$ is the bulk material resistivity of an insulator with thickness $d$ on top of a conductor and $R_s$ is the surface resistivity. We take the $x$ and $y$ directions to be parallel to the surface. The resistance varies by several orders of magnitude, depending on temperature in orbit. It also changes due to radiation dose.

By integrating Eq. (4.3) with respect to time, we can calculate how the differential voltage builds up between the insulator surface and the exposed conductor. When the secondary electron yield $\gamma_{\text{ee}}$ is larger than unity and the bulk and surface resistance is large, the differential voltage can develop as schematically shown in Fig. 4.3. Approximating the insulator by a planar capacitor with 100-$\mu$m thickness and the relative dielectric constant of 3, the capacitance is 300 nF/m$^2$. Assuming the total current density is 10 $\mu$A/m$^2$, the right-hand side of Eq. (4.3) is 33 V/s. The pace of differential voltage buildup is much slower than the variation of satellite potential $\phi_{\text{sc}}$.

To calculate the temporal variation of the satellite potential and the differential voltage correctly, we need a three-dimensional charging simulation software such as NASCAP-2K [8, 9], SPIS [10], or MUSCAT [11]. In the three-dimensional simulation, we need to take into account the three-dimensional spacecraft

**FIG. 4.3  Schematic example of differential charging buildup in GEO.**

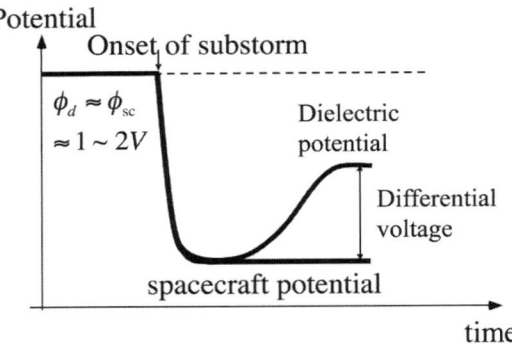

geometry, spacecraft surface material properties, and the temporal variation of all of the current densities.

During an eclipse, NPG or IPG can occur on many parts of the insulator surface. As the spacecraft exits eclipse and its solar array begins to generate electricity, there might still be large differential voltages left on the solar array. Because the solar-array voltage is high just after the eclipse due to low temperature, we should be careful to avoid occurrence of primary arcs and possible subsequent sustained arcs on the solar array just after the eclipse, as will be discussed in detail later.

## 4.3.2  SURFACE DIFFERENTIAL CHARGING IN LEO WITH A LOW INCLINATION ANGLE

The plasma density in LEO is larger than that in GEO by four to six orders of magnitude. Therefore, the dominant current sources into the spacecraft body are ionospheric electrons with temperatures of 0.1 to 0.2 eV and ram ions due to the spacecraft velocity of $\sim$7.6 km/s. Although the spacecraft is supersonic with respect to the ions, it is highly subsonic to the lighter electrons. The spacecraft body potential is determined to balance the fluxes of ionospheric electrons and ions. Because a conventional solar-array design has an exposed metallic surface such as interconnector, both the positive and negative ends of solar-array circuits are exposed to space. Therefore, the situation is similar to a floating double probe in a dense plasma, as shown in Fig. 4.4. The floating potential of the spacecraft is determined such that the electron current collected by the positive side is equal to the ion current collected by the negative side. The negative end of the solar-array circuit is connected to the spacecraft body, and conductive areas on the body surface also contribute to the collection of ions. Although

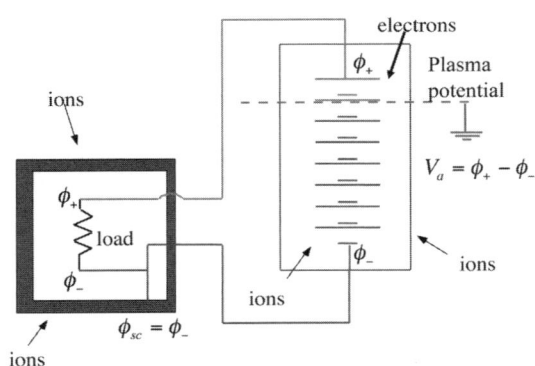

**FIG. 4.4  Current collection via a spacecraft in LEO.**

the negative side has a large collecting area, the mobility of electrons is far more than that of ionospheric ions. Therefore, most of solar-array generating voltage must become negative with respect to the plasma potential. As a rule of thumb, the spacecraft body potential $\phi_{sc}$ in LEO is approximated by $\phi_{sc} \approx -0.9V_a$ as a first-order approximation, where $V_a$ is the solar-array generating voltage.

The insulator surface potential reaches the steady state by attaining the zero net current condition locally. It is either a negative potential of the order of electron temperature or a positive potential of the order of the ion kinetic energy, ~5 eV. The positive potential appears where a negative sheath from a nearby exposed conductor shields the insulator surface [12]. Regardless of whether it is positive or negative, the values are negligible compared to the solar-array generation voltage $V_a$. Therefore, we can approximate the insulator surface potential by zero as a first-order approximation and approximate the average electric field at the triple junction by $V_a/d$. Because the insulator potential is more positive than the spacecraft potential, the IPG situation is the nominal condition in LEO during power generation.

### 4.3.3   SURFACE DIFFERENTIAL CHARGING IN LEO WITH A HIGH ORBITAL INCLINATION

When a spacecraft flies through the auroral zone, the dominant charged particle flux to the spacecraft body can be energetic auroral electrons. Negative charging of a polar orbiting satellite was observed for Freja and DMSP [13, 14] when the background ionospheric ion density was very low, $10^9$ m$^{-3}$ or less. Therefore, the satellite altitude must be sufficiently high, or the ion density must be abnormally low, for a satellite to experience negative charging. When negative charging occurs, either an IPG or NPG situation occurs at the triple junctions on the spacecraft exterior.

Severe differential charging is possible in the spacecraft wake, even if the spacecraft altitude is low enough not to have negative charging of spacecraft body (absolute charging). Over the Poles, the spacecraft solar paddle usually faces perpendicular to the flight direction. As the large spacecraft structure sweeps through the ionospheric plasma, ions cannot easily enter the wake region because the ion motion as seen from the spacecraft is supersonic. The energetic auroral electrons can reach the wake region easily and charge the insulator surface. At the ram side the spacecraft body can collect enough ions to keep the spacecraft body potential to $\phi_{sc} \approx -V_a$. Therefore, a NPG situation can occur in the wake side. According to two-dimensional numerical simulations [15, 16], a potential difference exceeding 1 kV is possible.

## 4.4   INVERTED POTENTIAL GRADIENT ESD INCEPTION MECHANISM

In this section, we describe the ESD inception mechanism under an IPG situation. Although we describe the ESD inception mechanism taking a solar array as the

**FIG. 4.5   Schematics of structure and cross section of a conventionally designed solar array.**

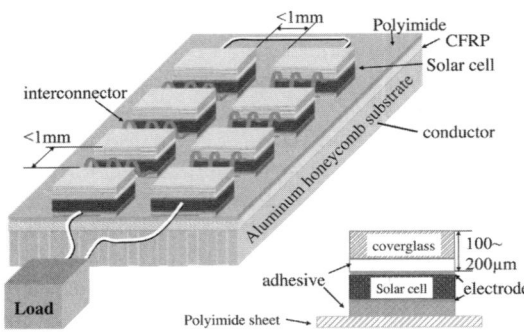

example in this section, the mechanism is the same for any other parts of a space-craft exterior that forms a triple junction.

A solar array generates a specified voltage by connect-ing solar cells in series. According to conventional designs of power circuits, the negative end of solar array is connected to the spacecraft body ground. In Fig. 4.5 the solar paddle structure, made of a conductor, usually carbon fiber reinforced plastic (CFRP) and aluminum honeycomb, has the same potential as the spacecraft body. The solar cells and interconnectors have a potential that is given by adding the voltage dependence on the position of each cell from the negative end of a solar-array circuit to the spacecraft body potential. A conventional design of the solar array has solar cells glued onto a polymer (usually polyimide) sheet over a CFRP/Al substrate. The coverglass is attached to a solar cell that has P and N elec-trodes. A metallic interconnector connects the P and N electrodes of adjacent solar cells in the string. Conventional designs expose the interconnector in order to absorb stresses induced by thermal cycling in orbit. The distance between solar cells is usually less than 1 mm. The thickness of the insulator on top of a solar cell is typically 100 to 200 μm, counting both the coverglass and a transparent adhesive.

It has been known since the 1970s [17, 18] that ESD occurs on a solar array under the IPG condition. There were many studies of this phenomenon from the second half of the 1980s to the first half of the 1990s. ESD studied then was a phenomenon that accompanied a pulse current event whose duration was several hundred nanoseconds to several microseconds long and was often called "arcing." To distinguish this discharge phenomenon from the sustained arc described later, this arc is nowadays often called a primary arc, primary ESD, or trigger arc.

We show a schematic of primary arc inception mechanism under the IPG situation proposed by [19] and [20] in Fig. 4.6. For the case of a solar array, the conductor in Fig. 4.6 corresponds to the interconnector, solar-cell electrode, or bus-bar, and the insulator in Fig. 4.6 corresponds to the coverglass, adhesive, or polyimide sheet. In Fig. 4.7, we show primary arc locations identified in a lab-oratory experiment [21], which clearly show that primary arcs are associated with the triple junctions.

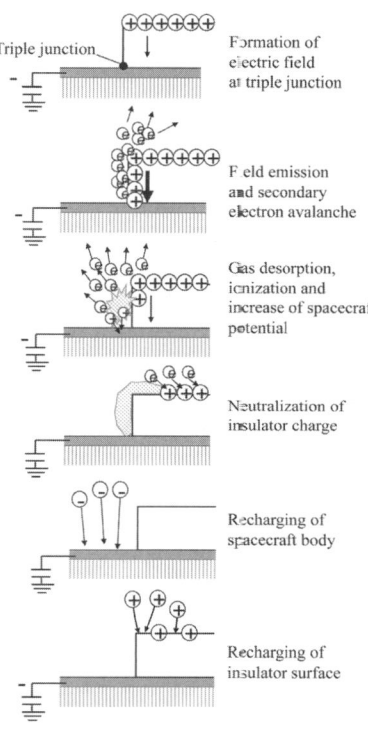

Formation of
electric field
at triple junction

Field emission
and secondary
electron avalanche

Gas desorption,
ionization and
increase of spacecraft
potential

Neutralization of
insulator charge

Recharging of
spacecraft body

Recharging of
insulator surface

**FIG. 4.6    Schematic picture of primary arc inception under an IPG situation.**

The scenario for arcing in an IPG situation is as follows:

1.  The insulator surface is positively charged. The arrow in Fig. 4.6 shows the direction of the electric field and illustrates that the field is intensified at the triple junction as the insulator surface accumulates positive charge. Positive charging of the insulator is done by ambient ions in LEO and secondary electrons or photoelectrons in GEO.

2.  As the electric field is further intensified, electrons are emitted from the conductor surface due to field emission. The potential structure around the triple junction produces an electric field where field-emitted electrons are attracted to the insulator surface. The electrons incident on the insulator surface emit secondary electrons that leave a positive charge near the triple junction and enhance the electric field further. The field emission electron current increases exponentially because of the feedback mechanism.

3.  As the field emission current increases, so does the electron current incident on the insulator. Then neutral gas is desorbed from the insulator surface as a result

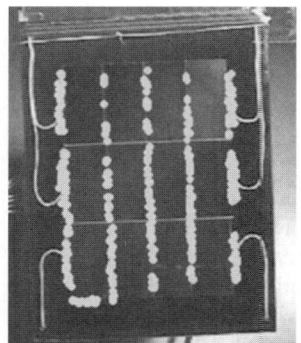

of the electron bombardment and forms a thin layer of neutral gas. The discharge occurs when the neutral gas is ionized. The discharge inception mechanism described here is similar to the theory of vacuum surface flashover [22, 23] although there is no anode in the present case.

**FIG. 4.7    Primary arc locations (circles) on solar array identified in a laboratory experiment under an IPG situation. The solar-array coupon was biased up to −500 V in a LEO-like plasma. Gaps between strings were filled with RTV Si rubber.**

4. As the discharge begins, positive charge on the insulator surface flows toward the conductor surface, forming the discharge current. Excessive heat melts the conductor surface, and the discharge becomes very similar to a vacuum arc where metallic vapor is the source of ionization, along with neutral gas desorbed from nearby surface. Craters found on the conductor surface [24] and the optical emission spectrum of gas molecules [25] and of metal vapor [26] support this picture.

5. As electrons escape toward ambient space from the discharge plasma, the charge stored on the capacitance between the spacecraft body and the ambient space is quickly discharged. This flow of charge is the blowoff current. Then, the spacecraft body potential rapidly increases to near zero. Because of capacitive coupling between the spacecraft body and the insulator surface, the insulator surface becomes even more positive of the ambient plasma.

6. The primary arc plasma neutralizes positive charge on the insulator surface via expansion of the surface flashover. This is the flashover current. In a dense LEO plasma, neutralization also occurs by collection of electrons from the ambient plasma. The primary arc or ESD ends when the current from the surface flashover (and, in LEO, the ambient plasma electron current) decreases below the current necessary to sustain the discharge.

7. The spacecraft body potential becomes negative again as the body acquires negative charges from the ambient plasma.

8. The insulator surface reacquires positive charges from the ambient plasma. As the surface charging proceeds, the situation goes back to the first step and repeats the process again.

When a primary arc or ESD occurs, the average electric field at the triple junction is of the order of megavolts per meter. In a macroscopic view this yields only an insignificant emission current. Therefore, there needs to be some mechanism to enhance the electric field microscopically, such as metallic whiskers or a dielectric impurity, by a factor of 100 to 1000. There are other proposed mechanisms for primary arc inception such as charging of a dielectric impurity on metallic surface via ambient ions [27].

The threshold values of primary arc inception on solar array in LEO plasma have been reported as 200 V for $V_a$ [28], corresponding to an electric field of megavolts per meter. One reason that the International Space Station solar-array voltage was chosen to be 160 V was to keep the voltage below this threshold. A space experiment carried out later observed arcing at a voltage as low as 75 V [29]. Recent laboratory studies on solar-array ESD have revealed that the threshold for primary arc inception is sometimes as low as 75 V [30, 31] in a LEO-like environment where a low temperature (typically $<< 1$ eV) and dense (typically $10^{10}$ to $10^{13}$ m$^{-3}$) rare gas plasma exist under a neutral background pressure of $10^{-3}$ to $10^{-2}$ Pa. A primary arc can occur wherever an exposed triple junction is formed. Examples are a CFRP surface, holes on a thermal insulator, exposure on an underlying conductive layer, and a crack in the insulation of a power cable. In particular, a primary arc can occur very easily on a CFRP surface where conductive carbon fibers are mixed with dielectric resin, exposing numerous triple junctions. Arcs on a CFRP surface were observed in laboratory tests even with a potential less than 100 V [32, 33].

On the other hand, in a GEO-like laboratory environment (where we use a beam of electrons with kilo-electron-volt energy and 10 to 1000 $\mu A/m^2$ current density to charge a solar-array test coupon under neutral background pressure of $10^{-4}$ to $10^{-3}$ Pa) primary arcs occur when the differential voltage exceeds 400 V or higher [34]. The differential voltage threshold is clearly higher for a GEO-like environment than for a LEO-like environment. Whether it is due to the difference of the background pressure, the difference of the methods of charging, or the dearth of ambient ions [35] is not clear for the present moment.

## 4.5  PRIMARY ARC GROWTH AND ITS DETRIMENTAL EFFECTS

A primary arc starts as a localized ionization phenomenon, but can evolve into a phenomenon that involves an entire solar paddle, or even the whole spacecraft structure. The primary arc plasma immediately after the arc inception grows by receiving electrostatic energy stored in the absolute capacitance. At the primary arc inception, positive charges flow into the spacecraft body through the primary arc location on the conductor surface. Because a typical spacecraft capacitance to space (its absolute capacitance) is less than 1 nF, the spacecraft body potential rapidly increases toward zero. The electrostatic energy stored on the absolute capacitance, more or less 1 $\mu J$, is given to the primary arc plasma during that process.

After the primary arc plasma uses the energy from the absolute capacitance, it relies on the electrostatic energy stored on nearby insulator surfaces. As the spacecraft potential approaches zero after the primary arc inception, the insulator surface potential has a positive value equal to the differential voltage before the primary arc inception (because of the aforementioned capacitive coupling). Then, the insulator surface attracts electrons in the primary arc plasma or the ambient plasma and releases its positive charge. The current path that neutralizes the positive charge on the insulator is now formed between the primary arc location and the insulator surface through the primary arc plasma and the spacecraft circuit. In Fig. 4.8, we show images of surface flashover propagation taken by a high-speed camera during a ground experiment [36]. At the same time we could measure the current from each string of solar array, the neutralization current. The current provided by the insulator of each string matched the arrival time of the flashover plasma.

Because the total capacitance of the spacecraft insulator surface often exceeds 10 $\mu F$, the total energy available for the primary arc plasma is as large as 10 J or even higher. How much of the insulator charge is neutralized by one primary arc is still unknown. Because it is very difficult to characterize the surface flashover

**FIG. 4.8** Propagation of flashover taken by a high-speed video camera at 500,000 frame/s with exposure time 1 μs. The circle in the left frame indicates the primary arc location. The middle frame is 1.1 μs, and the right frame is 3.0 μs from a primary arc inception, respectively.

propagation by on-orbit observation, we have to rely on a laboratory experiment. Currently, the maximum size of a solar array tested on the ground is 1.8 m × 1.1 m, and the surface flashover was observed to sometimes extend over the entire surface [37]. The present best estimate of the propagation speed of surface flashover is 10 km/s for a GEO solar array under an inverted potential gradient. (See [1, 36, and 38] for examples of the current waveform derivation.)

A primary arc on a solar array can cause various detrimental effects to the solar-array operation. The large transient current can cause electromagnetic interference with spacecraft circuits. As primary arcs can occur repeatedly during the lifetime of spacecraft, there can be cumulative effects such as surface degradation, surface contamination via vaporized material, or internal damage to solar cells due to surge voltages. When a primary arc occurs at the edge of a solar cell or bypass diode, it can degrade the electrical performance.

a)

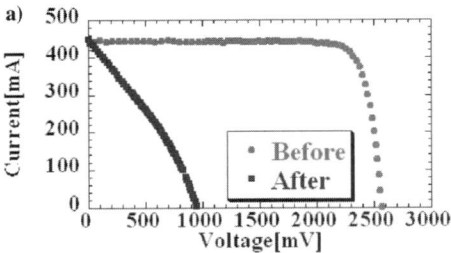

**FIG. 4.9** Example of solar-cell electrical output degradation due to primary arcs obtained in a) a ground experiment. Three primary arcs of average energy of 24 mJ were observed on an InGaP/GaAs/Ge triple junction solar-cell edge. Figure b) shows a photograph of a primary arc spot taken by a microscope after the experiment.

b)

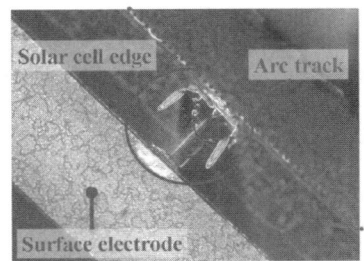

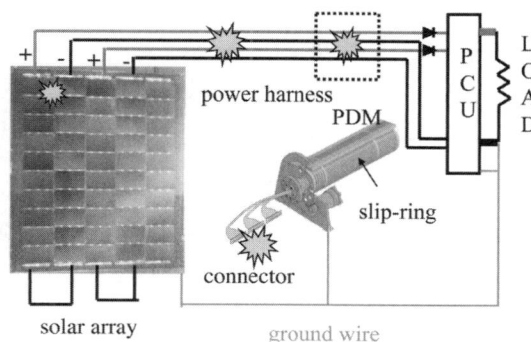

**FIG. 4.10   Typical spacecraft power system and points with secondary arc risk.**

Figure 4.9 shows an example of electrical performance degradation observed in a ground experiment [39].

## 4.6   SECONDARY ARC ON THE SPACECRAFT POWER SYSTEM

A primary arc alone does not usually cause fatal damage to spacecraft operation, although cumulative effects must be taken into account in spacecraft design. It can, however, trigger a secondary arc that sometimes produces fatal damage to a spacecraft power system. In Fig. 4.10 we show a schematic diagram of a spacecraft power system. When a primary arc occurs at points marked in the figure, there is a possibility that primary arc plasma might short circuit the power system.

Figure 4.11 schematically explains the growth of the primary arc plasma and the secondary arc formation. If the primary arc occurs at a gap between two points with a large potential difference within a solar-array string circuit, the arc plasma can short circuit the two points. For a conventionally designed solar paddle, the solar-array string layout has a serpentine shape in order to maximize the number of solar cells in a limited area and to suppress the magnetic moment induced by the string current. Therefore, it is not uncommon that the positive and negative ends of a solar-array string are adjacent to each other. A typical gap distance between solar cells is less than 1 mm. Across such a narrow gap, there is a voltage difference as large as the solar-array output voltage. One string of a solar-array circuit nowadays provides a current of several amperes. If the arc plasma resistance is sufficiently low, it can steal the solar-array power that might have otherwise gone to the spacecraft load. This stage of the arc is called a secondary arc or sustained arc.

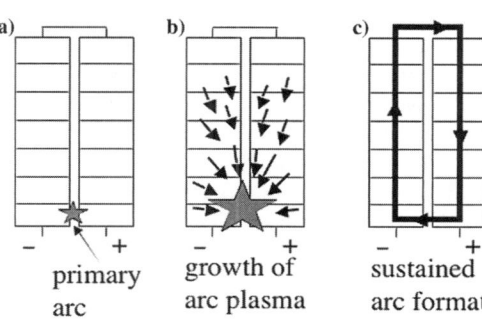

**FIG. 4.11   Schematic of secondary arc formation on solar array.**

**FIG. 4.12 Terminology related to the secondary arc phenomena. The current $I_{sc}$ is the short-circuit current that is available from the solar array.**

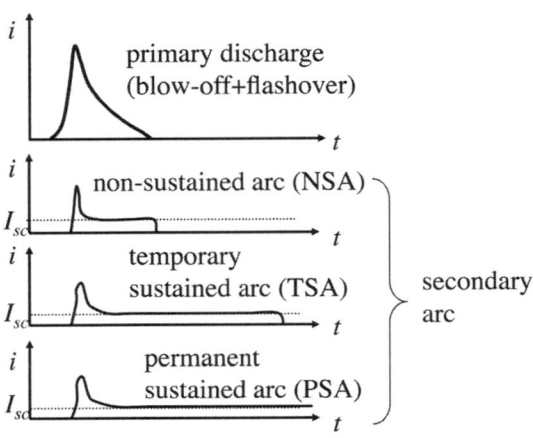

Figure 4.12 defines the terminology used in this text regarding the solar-array secondary arc. "Primary discharge" is the general term of ESD that includes both IPG ESD (primary arc) and NPG ESD. The primary discharge is fed by the absolute and differential capacitances. The primary discharge can create a temporary conductive path between any two cells with different voltages generated by solar-array power generation. The current between the two cells can become sustained when fed from the solar-array output power. This resulting phenomenon is called a "secondary arc." The secondary arc is powered by an illuminated solar array and leads to an important additional damage mechanism because the solar-array circuit can become permanently short circuited.

The secondary arc might or might not be self-sustained. There are several stages of a secondary arc. The primary discharge current is a sum of the blowoff current due to the absolute capacitance and the flashover current due to the differential capacitance (as well as the electron collection from the ambient plasma in a LEO environment). The secondary arc current is the current flowing between the two points through the illuminated solar array. In Fig. 4.12, $I_{sc}$ represents the short-circuit current of one or more solar-array circuits. A nonsustained arc (NSA) is a secondary arc that lasts only during the primary discharge. It ends when the primary discharge stops. A self-sustained arc is a secondary arc that lasts longer than the primary discharge. A temporary sustained arc (TSA) is self-sustained but stops by itself even though the solar-array power is still available. A permanent sustained arc (PSA) is an arc that does not stop as long as the solar-array power is available. A PSA can last for seconds or minutes. The very intense heat in the vicinity of the secondary arc creates a highly conductive path by pyrolizing materials around the discharge site. When a solar array is laid down on an insulator face-sheet, with typically a 50-$\mu$m thickness over a conductive substrate, a pyrolized insulator leaves a permanent conductive path even after the external power is shut off. The solar-array circuit then remains short circuited to the solar panel ground, diverting the solar-array power to the panel ground as shown in Fig. 4.13.

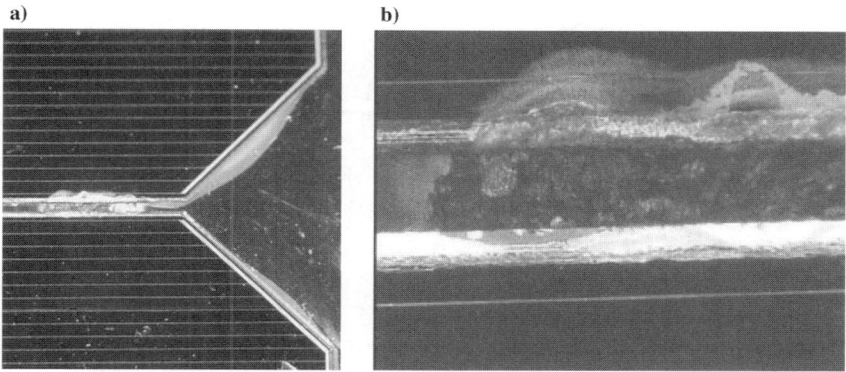

**FIG. 4.13    Photographs a) of solar cells damaged by a permanent sustained arc and b) 60x magnification of the insulation damage. The gap distance is approximately 600 μm.**

Figure 4.14 shows an example of TSA waveforms observed in a laboratory experiment [40]. The voltage across the gap (string voltage) dropped from 100 V to less than 50 V once the TSA started. The arc resistance calculated from the waveform is nearly 30 Ω. A variety of laboratory data [41–44] now exists to prove that a permanent sustained arc can occur with a voltage of 50 V or more and a current of 0.5 A or more for a gap distance of 1 mm or less. The temporary or permanent sustained arc is one type of vacuum surface arc, though the current level of 0.5 A for GaAs and 1.3 A for Si solar cell is smaller than the currents seen for a conventional vacuum arc, typically 10 A or higher [45]. How long the TSA continues depends on many parameters such as gap distance, current, voltage, gap insulator material, energy injected into the primary arc plasma, and others. A recent study [44] shows the TSA duration depends on the arc current available rather more strongly than the gap voltage, charging environment, or the primary arc energy, as shown in Fig. 4.15 (see also [46] for similar data).

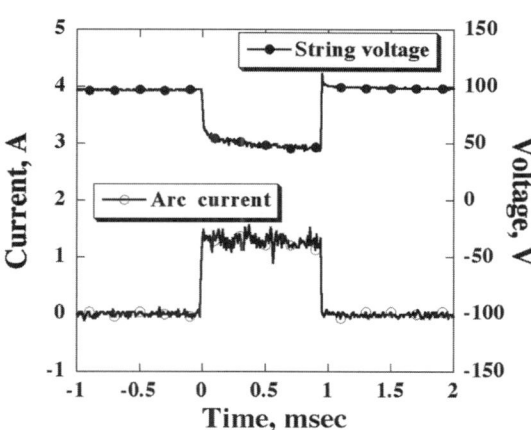

**FIG. 4.14    Example of TSA waveforms observed at a gap of 690 μm between silicon cells in low-temperature plasma environment.**

FIG. 4.15 Average TSA duration for various string currents, string voltages, and primary arc energies. Gaps of 1-mm distance between InGaP/GaAs/Ge solar cells were tested in a laboratory using an energetic electron beam as the charging method.

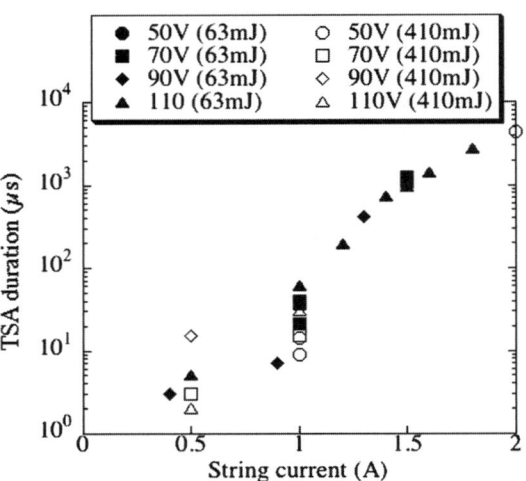

The sustained arc was first proposed as a failure mechanism for the Tempo-2 solar array [47]. Since Tempo-2, many satellites have suffered power system failures, and the prime suspects have been sustained arcs. In 2003, a Japanese Earth observation satellite, ADEOS-II (Midori-II) experienced a sudden power drop from 6 to 1 kW, leading to the complete loss of the satellite. The failure investigation [48] revealed that a sustained arc on the power harness at the solar paddle boom, triggered by a primary discharge between the inner layer of its thermal blanket and a wire crack, was responsible for the power drop. Energetic electrons that accompanied an extremely large auroral event in October 2003 had charged the ungrounded thermal blanket. In addition to solar arrays and power harnesses, a sustained arc at a solar paddle drive motor has also been the target of the failure investigation for a GEO satellite [49, 50].

There have been several mitigation techniques against sustained arcs [51, 52]:

1. Arrange the solar-array circuit layout to keep the maximum voltage between adjacent solar cells as low as possible.

2. Avoid the buildup of differential charging by increasing the conductivity of insulators or by the use of a plasma contactor.

3. Fill the gap between solar-array strings with the RTV silicon rubber that is used to glue the solar cells to the polymer sheet.

4. Insert a blocking diode at the positive end of solar-array strings to stop the current from flowing from other nonarcing strings even after the positive end of one string is short circuited to the spacecraft ground.

5. Insert a large resistance between the solar paddle structure and the spacecraft ground to make the solar-array current flow toward the spacecraft load even after the string is short circuited to the paddle structure.

Even though mitigation techniques have been utilized, many commercial satellites still fail in orbit. There is increasing demand for thorough ground tests before launch to prove that a given design of solar array will not suffer permanent sustained arc in orbit. Because the sustained-arc phenomenon has been recognized only recently, there is not yet a standardized ground-test method, although efforts have been intiated on an International Standard Organization (ISO) standard. As recent commercial satellites involve increasing numbers of international entities, ground tests based on an international standard are desirable. There are, however, still several discrepancies between different countries regarding how we should best simulate the reality in orbit.

There are always limitations in simulating the real space environment in a vacuum chamber of a finite size. One example is how we simulate the primary arc current that in orbit can be provided by the coverglasses on a solar paddle exceeding 10 m in length. Because the vacuum chamber size is limited, we cannot easily put a whole solar paddle in a chamber. If the primary arc current provided by the coverglasses far from the arc location is important to the occurrence of sustained arc, we have to find a way to simulate the current waveform via an external circuit. An international collaboration project [53, 54] has been started to establish an ISO standard for solar-array ESD testing by understanding the real physics involved in the arc phenomena and proposing the best laboratory simulation methods that are also affordable.

## 4.7 DISCHARGE CAUSED BY INTERNAL CHARGING

In addition to surface charging, there is a problem of internal charging caused by energetic charged particles in medium Earth orbit (MEO), GEO, geostationary transfer orbit (GTO), or a planetary radiation belt such as Jupiter. There are two cases of internal charging. In one case, the particles go through spacecraft exterior and charge dielectric materials or floating metals inside a spacecraft body. In another case, the charged particles can penetrate into the dielectric material on the spacecraft exterior, such as on cables and thermal insulation, and charge the dielectric material. This type of charging is also called *deep-dielectric charging*. Electrons with energies higher than 100 keV are the cause of these phenomena. Because the penetration depth of an ion is less than that of electrons by one or two orders of magnitude, ions are relatively unimportant for internal charging. A typical spacecraft exterior can be made of aluminum with a thickness of 750 μm or more, so that an electron with energy higher than ~500 keV can go through that thickness [55]. Even if the energy is less than 500 keV, electrons can stay inside exterior dielectric materials and leave their electrical charge. The electric field induced by internal charging depends on the material's dielectric constant and conductivity and the electron flux. When the electric field exceeds the threshold value, dielectric breakdown occurs, and a discharge current flows. The discharge as a result of internal charging occurs at a point closer to the spacecraft internal circuit as compared to the discharge caused by surface charging and thereby can lead to more serious effects on spacecraft operation.

According to [56], anomalies as a result of internal charging make up a majority of the past spacecraft anomalies. On 20 January 1994, three geosynchronous satellites (Intelsat-K, Anik E-1, and Anik E-2) suffered anomalies. In the case of Anik E-2, it took half a year before full functionality was restored. Considering the price of the satellite (several hundred million dollars), the financial loss caused by the service interruption was enormous [57, 58].

Internal charging occurs when electrons of higher than 100-keV energy appear as a result of the onset of a geomagnetic storm, irrespective of satellite position. According to data taken by CRRES [59], discharge occurs when $>10^5$ electrons accumulate over each 1 $cm^2$ area per second continuously for 10 h or more [60, 61]. These numbers are derived from ground tests and space experiment data from satellites such as CRRES or SCATHA. Theses numbers are not, however, based on measurement of the actual charge accumulated inside the dielectric materials. Therefore, technology is needed to measure the charging inside dielectric material [62]. Because internal charging is strongly correlated with magnetic storms, efforts to predict the emergence of magnetic storms from space weather forecasts, along with a system to announce proper warning to satellite operators, are also necessary.

## 4.8   CONCLUSION

Preventing ESD on spacecraft has been the central subject of spacecraft charging study for the past 40 years. The mechanism of surface charging and subsequent discharge has been studied by various space experiments as well as laboratory experiments and theoretical/computational analyses. Nowadays, fair consensus exists about how a primary discharge occurs on spacecraft surface under IPG. What follows the primary discharge, surface flashover, and secondary arc are also well known. Based on this knowledge, laboratory tests are carried out to qualify the high-voltage bus technologies. To improve the quantitative accuracy of the laboratory tests, detailed physical processes of surface flashover and secondary arc need to be explored to provide a mean to quantitatively assess the test data.

Various satellite failure investigations have identified secondary arc as the cause of failure. Yet, concrete evidence of surface flashover and secondary arc based on deliberate measurement and observation in orbit has not been obtained. Various techniques to mitigate primary ESD and spacecraft charging have been proposed so far. Although secondary arc mitigations, such as RTV gout and a bleeding resistor, were quick to be adapted by industries, few techniques to prevent the secondary arc at a more upstream stage, mitigation of ESD or spacecraft charging, have been demonstrated in orbit. To further improve the reliability of satellites, there continues a strong need of in-orbit study on ESD phenomena and its mitigation.

Internal electrostatic discharge has become increasingly important as space missions utilizing radiation-rich orbits have become more common and electronic circuit boards have become more susceptible to a discharge pulse. Quantifying charge accumulation and relaxation in dielectric materials bombarded by a realistic spectrum of energetic particles is necessary. This requires development of measurement method, characterization of conductivity including radiation-induced-conductivity, and validation of theoretical models via detailed laboratory experiments and in-orbit measurement.

# REFERENCES

[1]    Leung, P., "Plasma Phenomena Associated with Solar Array Discharges and Their Role in Scaling Coupon Test Results to a Full Panel," AIAA Paper 2002-0628, Jan. 2002.

[2]    Balmain, K. G., "Surface Discharge Effects," *Space Systems and Their Interactions with Earth's Space Environment*, edited by H. Garret and C. Pike, Progress in Astronautics and Aeronautics, Vol. 71, AIAA, Washington, D.C., 1980, pp. 276–298.

[3]    Ferguson, D. C., and Hillard, G. B., "Low Earth Orbit Spacecraft Charging Design Guidelines," NASA/TP-2003-212287, Feb. 2003.

[4]    Stevens, N. J., Mills, H. E., and Orange, L., "Voltage Gradients in Solar Array Cavities as Possible Breakdown Sites in Spacecraft-Charging-Induced Discharges," *IEEE Transactions on Nuclear Science*, Vol. 28, No. 6, 1981, pp. 4558–4562.

[5]    Levy, L., Rulet, R., Sarrail, D., Siguer, J. M., and Lechte, H., "MARECS and ECS Anomalies Attempt for Insulation Defect Production in Kapton," *Proceedings of the 5th European Symposium on Photovoltaic Generators in Space*, edited by W. R. Burke, Vol. 1, ESA, Paris, 1986, pp. 161–169.

[6]    Snyder, D. B., "Environmentally Induced Discharges in a Solar Array," *IEEE Transactions on Nuclear Science*, Vol. 29, No. 6, 1982, pp. 1607–1609.

[7]    Leung, P., "Discharge Characteristics of a Simulated Solar Cell Array," *IEEE Transactions on Nuclear Science*, Vol. 30, No. 6, 1983, pp. 4311–4315.

[8]    Mandell, M. J., Katz, I., Hilton, J. M., Minor, J., and Cooke, D. L., "NASCAP2K - A Spacecraft Charging Analysis Code for the 21st Century," AIAA Paper 2001-0957, Jan. 2001.

[9]    Davis, V. A., Neergaard, L. F., Mandell, M. J., Katz, I., Gardner, B. M., Hilton, J. M., and Minor, J., "Spacecraft Charging Calculations: NASCAP-2K and SEE Spacecraft Charging Handbook," AIAA Paper 2002-0626, Jan. 2002.

[10]   Roussel, J-F., Rogier, F., Dufour, G., Mateo-Velez, J-C., Forest, J., Hilgers, A., Rodgers, D., Girard, L., and Payan, D., "SPIS Open-Source Code: Methods, Capabilities, Achievements, and Prospects," *IEEE Transactions on Plasma Science*, Vol. 36, No. 5, Oct. 2008, pp. 2360–2368.

[11]   Muranaka, T., Hosoda, S., Kim, J., Hatta, S., Ikeda, K., Hamanaga, T., Cho, M., Usui, H., Ueda, O. H., Koga, K., and Goka, T., "Development of Multi-Utility Spacecraft Charging Analysis Tool (MUSCAT)," *IEEE Transactions on Plasma Science*, Vol. 36, No. 5, Oct. 2008, pp. 2336–2349.

[12]  Cho, M., and Hastings, D. E., "Dielectric Charging Processes and Arcing Rates of High Voltage Solar Arrays," *Journal of Spacecraft and Rockets*, Vol. 28, No. 6, 1991, pp. 698–706.

[13]  Anderson, P. C., "A Survey of Surface Charging Events on the DMSP Spacecraft in LEO," *Proceedings of 7th Spacecraft Charging Technology Conference* [CD-ROM], edited by E. Daly, SP476, ESA, Noordwijk, The Netherlands, 2001, pp. 331–336.

[14]  Wahlund, J. E., Wedin, L. J., Carrozi, T., Eriksson, A. I., Holback, B., Andersson, L., and Laakso, H., "Analysis of Freja Charging Events: Statistical Occurrence of Charging Events," ESA, ESTEC, Technical Note, SPEE-WP130-TN, Feb.1999.

[15]  Wang, J., Leng, P., Garret, H., and Murphy, G., "Multibody-Plasma Interactions: Charging in the Wake," *Journal of Spacecraft and Rockets*, Vol. 31, No. 5, 1994, pp. 889–894.

[16]  Cho, M., Kim, J., Hosoda, S., Miura, T., Iwata, T., and Nozaki, Y., "Electrostatic Discharge Ground Test of a Polar Orbit Satellite Solar Panel," *IEEE Transactions on Plasma Science*, Vol. 34, No. 5, 2006, pp. 2011–2030.

[17]  Kennerud, K. L., "High Voltage Solar Array Experiments," NASA CR-121280, March 1974.

[18]  Herron, B. G., Bayless, J. R., and Worden, J. D., "High Voltage Solar Array Technology," AIAA Paper 72-443, April 1972.

[19]  Hastings, D. E., Cho, M., and Kuninaka, H., "The Arcing Rate for a High Voltage Solar Array: Theory, Experiment and Predictions," *Journal of Spacecraft and Rockets*, Vol. 29, No. 4, 1992, pp. 538–554.

[20]  Hastings, D. E., Wyle, G., and Kaufman, D., "Threshold Voltage for Arcing on Negatively Biased Solar Arrays," *Journal of Spacecraft and Rockets*, Vol. 27, No. 5, 1990, pp. 539–544.

[21]  Hosoda, S., Okumura, T., Toyoda, K., and Cho, M., "High Voltage Solar Array for 400 V Operation in LEO Plasma Environment," 8th Spacecraft. Charging Tech. Conference [CD-ROM], edited by J. Minor, NASA Marshall Space Flight Center, Huntsville, AL, Oct. 2003.

[22]  Anderson, R. A., and Brainard, J. P., "Mechanism of Pulsed Surface Flashover Involving Electron-Stimulated Desorption," *Journal of Applied Physics*, Vol. 51, No. 3, 1980, pp. 1414–1421.

[23]  Pillai, A. S., and Hackam, R., "Modification of Electric Field at the Solid Insulator-Vacuum Interface Arising from Surface Charges on the Solid Insulator," *Journal of Applied Physics*, Vol. 54, No. 3, 1983, pp. 1302–1313.

[24]  Amorim, E., Levy, L., and Vacquie, S., "Electrostatic Discharges; Common Characteristics," *Journal of Physics D: Applied Physics*, Vol. 35, No. 7, 2002, pp. L21–L23.

[25]  Vayner, B., Galofaro, J., and Ferguson, D., "Interactions of High-Voltage Solar Arrays with Their Plasma Environment: Physical Processes," *Journal of Spacecraft and Rockets*, Vol. 41, No. 6, 2004, pp. 1031–1041.

[26]  Upschulte, B. L., Marinelli, W. J., Wyle, G. M., Aifer, E., and Hastings, D. E., "Arcing of Negatively Biased Solar Cells in a Plasma Environments," *Journal of Spacecraft and Rockets*, Vol. 31, No. 3, 1994, pp. 493–501.

[27]  Parks, D. E., Jongeward, G. A., Katz, I., and Davis, V. A., "Threshold-Determining Mechanism for Discharges in High-Voltage Solar Arrays," *Journal of Spacecraft and Rockets*, Vol. 24, No. 4, 1987, pp. 367–371.

[28]  Ferguson, D. C., "The Voltage Threshold for Arcing for Solar Cells in LEO-Flight and Ground Test Results," NASA TM-87259, March 1986.

[29]  Soldi, J. D., Hastings, D. E., Hardy, D., Guidice, D., and Ray, K.., "Flight Data Analysis for Photovoltaic Array Space Power Plus Diagnostis Experiment," *Journal of Spacecraft and Rockets*, Vol. 34, No. 1, 1997, pp. 92–103.

[30]  Vayner, B., Galofaro, J., and Ferguson, D., "Interactions of High-Voltage Solar Arrays with Their Plasma Environment: Ground Tests," *Journal of Spacecraft and Rockets*, Vol. 41, No. 6, 2004, pp. 1042–1050.

[31]  Davis, S., Stillwell, R., Andirario, W., Snyder, D., and Katz, I., "EOS-AM Solar Array Arc Mitigation Design," Society of Automotive Engineers, SAE Technical Paper Series, 1999-01-2582, Aug. 1999.

[32]  Hosoda, S., Okumura, T., Cho, M., and Toyoda, K., "Development of 400 V Solar Array Technology for Low Earth Orbit Plasma Environment," *IEEE Transactions on Plasma Science*, Vol. 34, No. 5, 2006, pp. 1986–1996.

[33]  Iwata, T., Miura, T., Nozaki, Y., Hosoda, S., and Cho, M., "Solar Array Paddle for the Advanced Land Observing Satellite (ALOS): Charging Mitigation and Verification," *Proceedings of the 9th Spacecraft Charging Technology Conference*, edited by T. Goka, Vol. 1, Japan Aerospace Exploration Agency, Tsukuba, Japan, 2005, pp. 840–856.

[34]  Cho, M., Ramasamy, R., Matsumoto, T., Toyoda, K., Nozaki, Y., and Takahashi, M., "Laboratory Tests on 110-Volt Solar Arrays in Simulated Geosynchronous Orbit Environment," *Journal of Spacecraft and Rockets*, Vol. 40, No. 2, 2003, pp. 211–220.

[35]  Vayner, B. V., Ferguson, D. C., and Galofaro, J. T., "Comparative Analysis of Arcing in LEO and GEO Simulated Environments," AIAA Paper 2007-93, Jan. 2007.

[36]  Masui, H., Toyoda, K., and Cho, M., "Electrostatic Discharge Plasma Propagation Velocity on Solar Panel in Simulated Geosynchronous Environment," *IEEE Transactions on Plasma Science*, Vol. 36, Oct. 2008, pp. 2387–2394.

[37]  Mashidori, H., Nitta, K., Kawakita, S., Shinohara, S., Harada, J., Miki, Y., and Toyoda, K., "Preliminary ESD Ground Tests Using a Large-Scale Solar Array," *10th Spacecraft Charging Technology Conference Proceedings* [CD-ROM], edited by D. Fourny-Delloye, CNES, Touuose, France, June 2007.

[38]  Amorim, E., Payan, D., Reulet, R., and Sarrail, D., "Electrostatic Discharges on a 1m² Solar Array Coupon Influence of the Energy Stored on Coverglass on Flashover Current," *9th Spacecraft Charging Technology Conference Proceedings*, edited by T. Goka, Japan Aerospace Exploration Agency, Tsukuba, Japan, 2005, pp. 331–344.

[39]  Okumura, T., Ninomiya, S., Masui, H., Toyoda, K., Cho, M., and Imaizumi, M., "Solar Cell Degradation due to ESD for International Standardization of Solar Array ESD Test," *10th Spacecraft Charging Technology Conference Proceedings* [CD-ROM], edited by D. Fourny-Delloye, CNES, Toulouse, France, 2007.

[40]  Aso, S., Kitamura, T., Kim, J., Hosoda, S., Cho, M., and Kagawa, H., "Threshold for Secondary Arc Formation on Solar Array in Low Earth Orbit Plasma Environment," *Transactions of the Japan Society Aero. Space Sciences*, Vol. 53, No. 622, 2005, pp. 516–523.

[41]  Cho, M., Ramasamy, R., Toyoda, K., Nozaki, Y., and Takahashi, M., "Laboratory Tests on 110-Volt Solar Arrays in Simulated Geosynchronous Orbit Environment," *Journal of Spacecraft and Rockets*, Vol. 40, No. 2, 2003, pp. 221–229.

[42]  Gaillot, L., Fille, M-L., and Levy, L., "Secondary Arcs on Solar Array – Test Results of EMAGS 2," *Proceedings of the 9th Spacecraft Charging Technology Conference*, edited

by T. Goka, Japan Aerospace Exploration Agency, Tsukuba, Japan, 2005, pp. 345–352.

[43]    Kim, J., Aso, S., Hosoda, S., and Cho, M., "Threshold Conditions to Induce the Sustained Arc on the Solar Array Panel of LEO Satellite," *Proceedings of the 9th Spacecraft Charging Technology Conference*, edited by T. Goka, Vol. 1, Japan Aerospace Exploration Agency, Tsukuba, Japan, 2005, pp. 179–191.

[44]    Kitamura, T., Masui, H., Toyoda, K., and Cho, M., "Secondary Arc Tests on Solar Arrays for International Standardization of ESD Test and Japanese Spacecraft Charging Guideline," *10th Spacecraft Charging Technology Conference Proceedings* [CD-ROM], edited by D. Fourny-Delloye, CNES, Toulouse, France, June 2007.

[45]    Boxman, R. L., Sanders, D. M., and Martin, P. J., *Handbook of Vacuum Arc Science and Technology Fundamentals and Applications*, 1st ed., Noyes Publications, Park Ridge, NJ, 1995, p. 119.

[46]    Okumura, T., Masui, H., Toyoda, K., Cho, M., Nitta, K., and Imaizumi, M., "Environmental Effects on Solar Array Electrostatic Discharge Current Waveforms and Test Results," *Journal of Spacecraft and Rockets*, Vol. 46, No. 3, 2009, pp. 697–705.

[47]    Katz, I., Davis, V. A., and Snder, D. B., "Mechanism for Spacecraft Charging Initiated Destruction of Solar Arrays in GEO," AIAA Paper 98-1002, Jan. 1998.

[48]    Maejima, H., Kawakita, S., Kusawake, H., Takahashi, M., Goka, T., Kurosaki, T., Nakamura, M., Toyoda, K., and Cho, M., "Investigation of Power System Failure of a LEO Satellite," AIAA Paper 2004-5657, Aug. 2004.

[49]    Amorim, E., Boulay, F., Migliorero, G., and Vielinguimert, V., "Arc Propagation on Space Power Transfer System: A First Approach Study," *Proceedings of the 9th Spacecraft Charging Technology Conference*, edited by T. Goka, Vol. 1, Japan Aerospace Exploration Agency, Tsukuba, Japan, 2005, pp. 530–537.

[50]    Taverna, M., "SES Global, Snecma Seek to Quell Space Market Unrest," *Aviation Week and Space Technology*, 9 Dec. 2002, p. 42.

[51]    Ferguson, D. C., Snyder, D. B., Vayner, B. V., and Galofaro, J. T., "Array Arcing in Orbit – From LEO to GEO," AIAA Paper 99-0218, Jan. 1999.

[52]    "Low Earth Orbit Spacecraft Charging Design Standard," NASA STD-4005, Sept. 2007.

[53]    Cho, M., and Goka, T., "Issues Associated with Standardization of Ground Test Methods of Electrostatic Discharge Phenomena on Spacecraft Surface," 56th International Astronautical Conference, Paper IAC-05-D5.1.03, Oct. 2005.

[54]    Cho, M., "Status of ISO Standardization Efforts of Solar Panel ESD Test Methods," *10th Spacecraft Charging Technology Conference Proceedings* [CD-ROM], edited by D. Fourny-Delloye, CNES, Toulouse, France, June 2007.

[55]    Garrett, H., and Whittlesey, A. C., "Spacecraft Charging an Update," *IEEE Transactions on Plasma Science*, Vol. 28, No. 6, 2000, pp. 2017–2028.

[56]    Koons, H. C., Mazur, J. E., Selesnick, R. S., Blake, J. B., Fennell, J. F., Roeder, J. L., and Anderson, P. C., "The Impact of the Space Environment on Space Systems," *Proceedings of the 6th Spacecraft Charging Technology Conference*, edited by D. L. Cooke, AFRPL-VS-TR-20001578, U.S. Air Force Research Lab., Hanscom AFB, MA, 2000, pp. 7–11.

[57]    Leach, R. D., and Alexander, M. B. (eds.), *Failures and Anomalies Attributed to Spacecraft Charging*, NASA Reference Publication 1375, NASA Marshall Space Flight Center, Huntsville, AL, 1995.

[58]   Baker, D. N., Kanekal, S., Blake, J. B., Klecker, B., and Rostoker, G., "Satellite
       Anomalies Linked to Electron Increase in the Magnetosphere, EOS," *Transactions of
       American Geophysical Union*, Vol. 75, No. 34, 1994, pp. 401–405.

[59]   Vampola, A. L., "Combined Release and Radiation Effects Satellite," *Journal of
       Spacecraft and Rockets*, Vol. 29, No. 4, 1992, p. 555.

[60]   Fennell, J. F., Koons, H. C., Chen, M. W., and Blake, J. B., "Internal Charging: A
       Preliminary Environmental Specification for Satellites," *IEEE Transactions on
       Plasma Science*, Vol. 28, No. 6, 2000, pp. 2029–2036.

[61]   Frederickson, A. R., Holeman, E. G., and Mullen, E. G., "Characteristics of
       Spontaneous Electrical Discharging of Various Insulator in Space Radiations," *IEEE
       Transactions on Nuclear Science*, Vol. 39, No. 6, 1992, pp. 1773–1782.

[62]   Takada, T., and Tanaka, Y., "Pulse Acoustic Technology for Measurement of Charge
       Distribution in Dielectric Materials for Spacecraft," *Proceedings of the 9th Spacecraft
       Charging Technology Conference*, edited by T. Goka, Vol. 1, Japan Aerospace
       Exploration Agency, Tsukuba, Japan, 2005, pp. 19–29.

# CHAPTER 5

# Spacecraft Charging Simulation

Mengu Cho*

*Kyushu Institute of Technology, Kitakyushu, Japan*

## 5.1   INTRODUCTION

Because there is no "electrical ground" in space, the ambient plasma surrounding a spacecraft serves as the reference point of its electrical potential. Although the plasma is a good conductor (its conductivity at LEO is comparable to or better than the seawater), the presence of low-mass electrons and high-mass ions usually makes the spacecraft potential with respect to the plasma $\phi_{sc}$ different from the plasma potential. As the density and temperature of the plasma always vary and depend on the orbit, the potential $\phi_{sc}$ is not a constant value. The spacecraft body structure serves as the reference point of the spacecraft circuit. Every electrical component is supposed to be grounded to the body structure. If a spacecraft were made of only conductive material, most of the spacecraft charging problems would have been avoided. As a large fraction of spacecraft surface is still nonconductive, however, the potential of the nonconductive surface $\phi_d$ is determined by its interaction with the surrounding plasma. The problems associated with absolute charging, that is, $\phi_{sc} \neq 0$, and differential charging, that is, $\phi_{sc} \neq \phi_d$, have been known since the early days of spaceflight. Many efforts have been made to predict $\phi_{sc}$ and $\phi_d$ using analytical formulas and numerical simulations. In this chapter, we will briefly look at the principle, availability, application, techniques, validation, improvement, and future of the spacecraft charging simulation.

## 5.2   PRINCIPLES OF SPACECRAFT CHARGING SIMULATION

The purpose of the spacecraft charging simulation is to find the spacecraft potential $\phi_{sc}$ for a given set of input parameters. The input parameters are generally the following:

- Ambient plasma environment such as density and temperature

---

*Director, Laboratory of Spacecraft Environment Interaction Engineering, 804-8550. Senior Member AIAA.

- Spacecraft position and attitude such as magnetic field and solar incident flux
- Spacecraft geometry such as size or even detailed three-dimensional geometry
- Spacecraft material data such as secondary electron coefficient, photoelectron coefficient, and conductivity

Finding a steady-state $\phi_{sc}$ is the first and minimum requirement. If only $\phi_{sc}$ is required, we often need to solve one equation for the charge balance of the spacecraft body.

$$I_e(\phi_{sc}) + I_i(\phi_{sc}) + I_{se}(\phi_{sc}) + I_{si}(\phi_{sc}) + I_{be}(\phi_{sc}) + I_{ph}(\phi_{sc}) + I_a(\phi_{sc}) + I_c(\phi_{sc}) = 0 \tag{5.1}$$

where $I_e$ is the ambient electron current, $I_i$ is the ambient ion, $I_{se}$ is the electron-induced secondary electron current, $I_{si}$ is the ion-induced secondary electron current, $I_{be}$ is the backscattered electron current, $I_{ph}$ is the photoelectron current, $I_a$ is the active emission current, and $I_c$ is the leak current from an insulator surface through the bulk or surface conductivity. Strictly speaking, the currents depend also on the insulator surface potential as it might deflect the charged particle trajectories, and the leakage current is determined by the difference between the insulator and the body potentials. If we consider a simple conductive body, the unknown is only $\phi_{sc}$ with one equation. By using any root-finding method, such as the Newton–Raphson method, the equation can be solved to find $\phi_{sc}$. If we can neglect all of the terms except the first and second terms, we can apply Langmuir probe theory and its expanded formula [1, 2] to calculate the potential $\phi_{sc}$.

The present sophistication of spacecraft design demands more than the simple estimate of the steady-state $\phi_{sc}$. The second requirement is to find the steady-state insulator potential $\phi_d$. Then, we have to solve the following set of equations:

$$\begin{aligned} \frac{d\phi_{sc}}{dt} &= \frac{1}{C_{sat}} [I_e(\phi_{sc}) + I_i(\phi_{sc}) + I_{se}(\phi_{sc}) + I_{si}(\phi_{sc}) + I_{be}(\phi_{sc}) \\ &\quad + I_{ph}(\phi_{sc}) + I_a(\phi_{sc}) + I_c(\phi_{sc})] \\ \frac{d(\phi_d - \phi_{sc})}{dt} &= \frac{1}{C_d} [j_e(\phi_d) + j_i(\phi_d) + j_{se}(\phi_d) + j_{si}(\phi_d) \\ &\quad + j_{be}(\phi_d) + j_{ph}(\phi_d) + j_c(\phi_d)] \end{aligned} \tag{5.2}$$

The problem here is that the insulator potential $\phi_d$ can be different at each point on a spacecraft. Therefore, strictly speaking, we have to solve a nearly infinite number of differential equations for $\phi_{sc}$ and $\phi_d$. Instead, we divide the spacecraft surface into a finite number of discrete elements and consider only a limited number of equations. Because the spacecraft surface insulator is usually very thin, typically less than 1 mm, the capacitance between the insulator surface and the underlying conductor that has the potential $\phi_{sc}$ dominates over the other

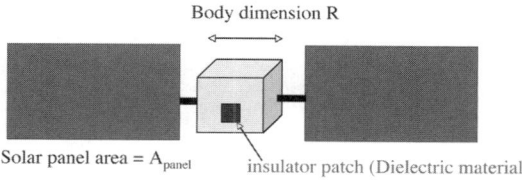

Body dimension R

Solar panel area = $A_{panel}$    insulator patch (Dielectric material)

FIG. 5.1   Generic spacecraft considered for a simple charging analysis.

capacitance between a point on the insulator and other points. Therefore, we can model the temporal change of the insulator surface potential as charging of the capacitance per unit area across the insulator $C_d$. Because the potential $\phi_{sc}$ is also defined as the potential difference between the ambient plasma and the spacecraft body, the temporal variation of $\phi_{sc}$ is also modeled as charging of the capacitance between the body and the plasma $C_{sat}$ by the current carried by the ambient plasma, secondary electrons, and others.

Let us simplify Eqs. (5.2) considering only three components as shown in Fig. 5.1. [Data for the QUSCAT (Quick Spacecraft Charging Analysis Tool) is available online at http://laplace.ele.kyutech.ac.jp/Charging/charging.html.] It is a generic three-axis-stabilized satellite. Its main body, whose total surface area is given by $A_{body}$, is covered by conductive material except a small patch of insulator whose area is given by $A_{patch}$. One side of the solar panel, whose area is given by $A_{panel}$, is covered by coverglass. The other side of the solar panel is covered by the same conductive material as the main body. For simplicity, we consider only the ambient electron current, the ambient ion current, the electron-induced secondary electron current, the photoelectron current, and the conduction current. Then Eqs. (5.2) can be simplified as

$$\frac{d\phi_{sc}}{dt} = \frac{1}{C_{sat}} \left\{ \begin{array}{l} [j_i(\phi_{sc}) + j_e(\phi_{sc}) + j_{se}(\phi_{sc})](A_{body} + A_{panel}) + j_{ph}(\phi_{sc})\dfrac{A_{body}}{2} \\[2mm] + [j_i(\phi_{cg}) + j_e(\phi_{cg}) + j_{se}(\phi_{cg}) + j_{ph}(\phi_{cg})]A_{panel} \end{array} \right\}$$

$$\frac{d\phi_{cg}}{dt} = \frac{1}{C_{cg}} [j_i(\phi_{cg}) + j_e(\phi_{cg}) + j_{se}(\phi_{cg}) + j_{ph}(\phi_{cg}) + j_c(\phi_{cg} - \phi_{sc})] + \frac{d\phi_{sc}}{dt}$$

$$\frac{d\phi_d}{dt} = \frac{1}{C_d} [j_i(\phi_d) + j_e(\phi_d) + j_{se}(\phi_d) + j_{ph}(\phi_d) + j_c(\phi_d - \phi_{sc})] + \frac{d\phi_{sc}}{dt}$$

$$(5.3)$$

where $\phi_{cg}$ and $\phi_d$ are the potentials of the solar-panel coverglass and the insulator patch, respectively. If we are required to give only the steady-state solutions, that is, $d\phi_{sc}/dt = d\phi_{cg}/dt = d\phi_d/dt = 0$, we solve the following three equations:

$$0 = [j_i(\phi_{sc}) + j_e(\phi_{sc}) + j_{se}(\phi_{sc})](A_{body} + A_{panel})$$

$$+ j_{ph}(\phi_{sc})\frac{A_{body}}{2} - [j_c(\phi_{cg} - \phi_{sc})]A_{panel}$$

$$0 = j_i(\phi_{cg}) + j_e(\phi_{cg}) + j_{se}(\phi_{cg}) + j_{ph}(\phi_{cg}) + j_c(\phi_{cg} - \phi_{sc})$$

$$0 = j_i(\phi_d) + j_e(\phi_d) + j_{se}(\phi_d) + j_{ph}(\phi_d) + j_c(\phi_d - \phi_{sc})$$

(5.4)

The first equation corresponds to Eq. (5.1) that is written in the form of total current $I(\phi_{sc}) = 0$. When it is in the steady state, the body potential is determined mostly by the current through the exposed conductor area, the main body, and the back of solar panel for the case shown in Fig. 5.1. During the transient stage, however, the charge balance across the entire spacecraft exterior surface determines the body potential. The second and third equations in Eqs. (5.4) give the steady-state insulator potential. Because the formulas for the current densities are nonlinear, there is no guarantee that there is a unique solution for Eqs. (5.4). There can be multiple sets of roots to satisfy Eqs. (5.4). In such a case, we have to integrate the temporal differential equations from an appropriate set of initial conditions.

The third requirement for the charging simulation is to provide a time necessary to charge the surface insulator to a certain voltage difference, $\Delta V = \phi d - \phi_{sc}$. This information is necessary to obtain the rate of discharge, which is given as the inverse of the charging time [3]. In Fig. 5.2 we show the temporal potential profiles obtained from MUSCAT simulation of a GEO satellite [4]. From a laboratory experiment, the threshold of electrostatic discharge (ESD) inception is known to be $\Delta V = 400(V)$ [5]. Figure 5.2 shows that the potential difference builds up in 40 s from the start of charging event for this particular set of conditions. After one ESD, the satellite potential jumps to near zero as the satellite capacitance is quickly discharged. The differential voltage $\Delta V$ is also nullified as the discharge plasma neutralizes the charge stored on the surface insulator. Then the charging situation returns the initial state. Therefore, if this plasma environment continues, it is expected that a satellite undergoes ESD once every 40 s. To obtain this type of temporal profile, we have to numerically integrate Eqs. (5.2) with respect to time. One difficulty is that the characteristic charging

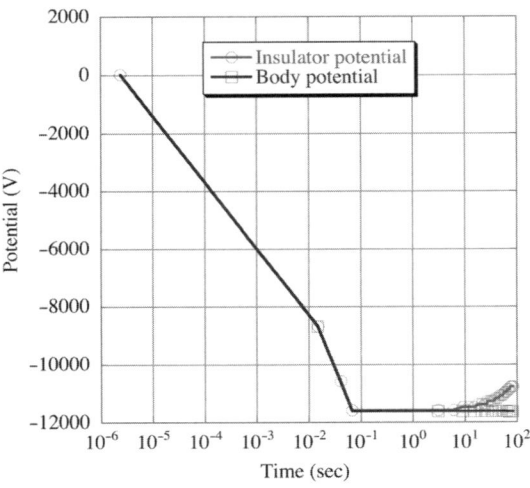

**FIG. 5.2 Example of charging profile calculated for a GEO satellite. (From Fig. 13 of [4]. © 2008 IEEE reprinted with permission.)**

time for the body is smaller than that of the insulator by many orders of magnitude. We need a numerical technique to overcome the difficulty. In Fig. 5.2, the time step of integration was varied during the simulation to solve the temporal profiles of both $\phi_{sc}$ and $\phi_d$.

The fourth requirement is to provide information on the plasma environment near a spacecraft as modified from the ambient conditions by the presence of the spacecraft. This requirement often comes from scientists who want to have a tool to know what they really measured as plasma in orbit. There are also practical reasons such as a study of the interaction between an electric propulsion thruster plume and a spacecraft body or a solar array [6]. We can fulfill the fourth requirement by retaining the information on the trajectories of charged particles and the space potentials around the spacecraft.

## 5.3 SPACECRAFT CHARGING SIMULATION CODES

The basic role of spacecraft charging simulation is to numerically integrate the differential equations (5.2) with respect to time for a finite number of surface elements on a spacecraft. We calculate the contribution of each current component in Eqs. (5.2) by following the trajectories of the ambient plasma particles, calculating the interaction between the incident particles and the surface material, and considering the emission of charged particles from the surface, such as secondary electrons, photoelectrons, and artificial plasmas.

Then, the surface charge distribution is updated based on the current density and conductivity of each surface element relative to the spacecraft body ground. The surface potentials are calculated by solving the Poisson equation $\nabla^2 \phi = -\rho/\varepsilon_o$ or the Laplace equation $\nabla^2 \phi = 0$. Particle trajectories are recalculated for the renewed surface potential, and the processes are repeated until the elapsed time reaches the predetermined time or the steady state is obtained.

Figure 5.3 illustrates a flowchart of a generic charging simulation code. An important feature of the practical charging simulation code is the capability to model the three-dimensional geometry of the spacecraft. Three-dimensional simulation is necessary for modern three-axis-stabilized spacecraft. Reference [7] shows "saddle point" charging in GEO where photoelectrons were blocked by the negative sheath from the nonsunlit surface leading to severe negative charging even in sunlight. Another important point is the capability to model surface interactions such as secondary electron and photoelectron emission. Having a reliable database of spacecraft surface material properties, such as conductivity, secondary electron emission yield, and photoelectron emission coefficient, is also important. There are several charging simulation codes available with these features as listed here:

*NASCAP-GEO* [8]: This NASA–USAF charging simulation code has been the most common spacecraft charging code in use by the space industry. It has been made available to various users via collaboration agreements with NASA or via

**FIG. 5.3 Generic flowchart of spacecraft charging simulation.**

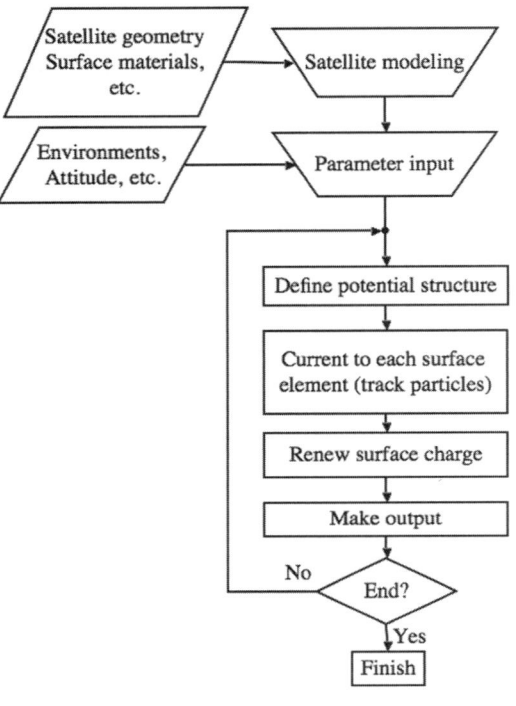

commercial arrangement. This code calculates the total current, due to all of the current contributions, for each surface on a numerically modeled three-dimensional spacecraft, using a two Maxwellian environment for both ions and electrons. From these currents, the change in potential at each surface is calculated. The current and potential calculations can be performed iteratively until an equilibrium charging state is achieved.

*NASCAP-LEO* [9]: NASCAP-LEO is aimed at the simulation of high potential objects with cold, dense plasma typical of the LEO environment. Like NASCAP-GEO, it employs analytical current collection equations, although these are aimed at sheath limited current collection and are appropriate for the short Debye-length LEO plasmas. Typical uses are the simulation of parasitic currents from high potential surfaces, such as solar-array interconnects.

*POLAR* [10]: The main problem with computing charging in LEO is computing efforts associated with the spacecraft sheath. The three-dimensional POLAR code has been designed for the assessment of sheath and wake effects on polar Earth orbit (PEO) spacecraft. POLAR uses numerical techniques to track ambient ions inwards from the electrostatic sheath surrounding a negatively charged spacecraft, onto the spacecraft surface. Spacecraft velocity is included as an input, and ram/wake effects are modeled. One or two Maxwellian components can be used to define the ambient plasma. The electron population in POLAR is a superposition of power-law, Maxwellian, and Gaussian components. Once the surface currents have been found, POLAR calculates potentials and the equilibrium charging state in a similar way to NASCAP.

*NASCAP-2K* [11, 12]: The most recent NASCAP code (NASCAP-2K) is available, free, to U.S. citizens only. This is a comprehensive code with realistic geometry. It combines the capabilities of NASCAP-GEO, NASCAP-LEO, and POLAR. The code is not available outside the United States.

*SPIS* [13, 14]: SPIS is a fully three-dimensional particle-in-cell (PIC) code that allows the exact computation of the sheath structure and the current collected by spacecraft surfaces for very detailed geometries. Surface interactions including photoelectron emission, backscattering, secondary-electron emission, and conduction are modeled. The source code is freely available from www.spis.org, and a mailing list provides a limited amount of support.

*MUSCAT* [4]: MUSCAT is a fully three-dimensional particle code that can be applied to spacecraft in LEO, PEO, and GEO. Its algorithm is a combination of PIC and particle tracking. A parallel computation technique is used for fast computation. It has a JAVA3D three-dimensional-based graphical user interface for three-dimensional modeling of spacecraft geometry and output visualization. The surface interactions included in the NASCAP series and SPIS are included as well. A material property database is also included. The code is commercially available.

*SPARCS* [15]: This is a code developed and used by Thales Alenia Space for simulating surface charging in the outer magnetosphere (GEO). This code computes the total current, due to all the primary and secondary current contributions, for each surface on a finite element type geometrical model of the spacecraft. Physical phenomena accounted for the secondary current budget are electron photoemission, backscattering and secondary cascade emission, and secondary electron recollection. SPARCS uses a two Maxwellian environment for both ions and electrons. From these currents, the change in the absolute potential of the spacecraft and the change in the differential potential at each surface are calculated. Environmental conditions can be changed during computation for a simulation of an end-of-eclipse. SPARCS can be made available commercially upon request.

*GES* [16]: Geospace environment simulation (GES) is a full-PIC code that uses a massive parallel supercomputer, the Earth Simulator. The Earth Simulator consists of 640 nodes of a vector-type supercomputer. Each node has 16-GB main memory shared with eight vector processors. The total size of the main memory is 10 TB, and the maximum performance is 40 trillion floating point operations per second (40 TFLOPS). The code is based on a PIC code developed by a group associated with Kyoto University. The parallel PIC code can solve electromagnetic interactions between plasma and a spacecraft body within a timescale of the electron plasma frequency. GES can import a spacecraft geometry file created by the graphical-user-interface (GUI) program of MUSCAT.

## 5.4    APPLICATIONS OF SPACECRAFT CHARGING SIMULATIONS

The most common application of spacecraft charging simulations is to study the differential charging of a GEO satellite under the worst-case environment. Based on SCATHA measurements, there is a recommended worst environment that describes the GEO plasma environment by a double Maxwellian [17]. When geometry and material specifications of a GEO satellite become available, the steady-state values of $\phi_{sc}$ and $\phi_d$ are calculated using the recommended double

FIG. 5.4    Example of a worst-case
charging environment simulation
by NASCAP-GEO. The spacecraft
body potential is −1450 V. Only
the differential voltage is shown.
The satellite is WINDS, a Japanese
three-axis-stabilized GEO satellite.

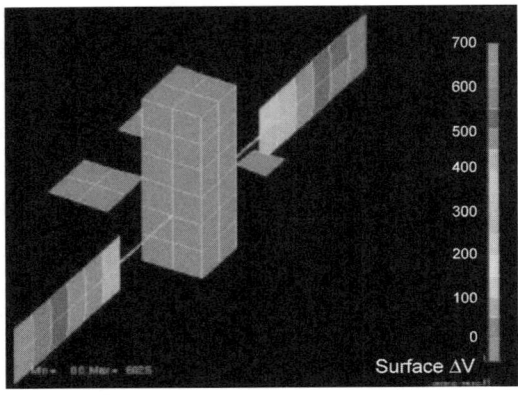

Maxwellian as the environmental parameters. NASCAP
has been used widely since
the end of 1970s for that
purpose. Figure 5.4 shows an
example of the charging potential calculated for WINDS, a Japanese GEO satellite.
Unless the coverglasses are conductively coated and grounded, serious differential
charging on the solar array cannot be avoided. Because of secondary electrons and
photoelectrons emitted from the coverglass, the coverglass becomes more positive
than the spacecraft body. The potential difference is usually the highest at the tip
of the solar paddle where the influence of the negative potential of the spacecraft
main body is least.

A second application of charging simulation codes is to derive proper experimental conditions. This application is divided into two stages. The first one is to
examine whether a spacecraft suffers ESD under the worst-case charging situation
derived in the first use. We create the same differential voltage on a test sample,
such as a solar-array coupon described in the chapter of surface discharge, as the
one predicted by the worst-case simulation using a laboratory facility. If we
observe no ESD even after we add enough margin to the simulation result, we
can declare the satellite is
free from ESD even under
the possible worst plasma
environment. The charg-
ing analysis result often
depends strongly on the
spacecraft material prop-
erties. Even if the worst-case

FIG. 5.5    Example of the
dependence of a NASCAP/GEO
simulation result on coverglass
conductivity. (From Fig. 10 of
[18]. Copyright © 2005 AIAA.
Reprinted with permission.)

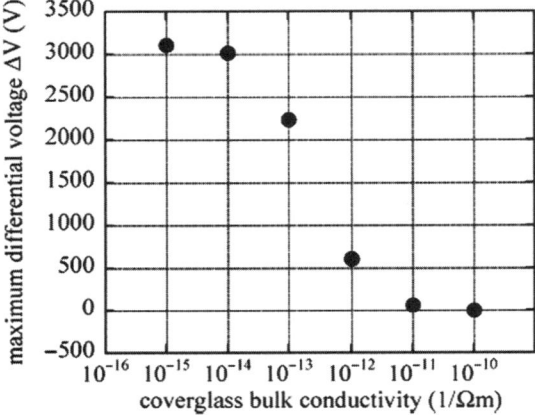

simulation shows no serious charging, the result should be confirmed by varying the material data within the possible range of the parameters. Figure 5.5 shows one example of such dependence. In this case we varied the material conductivity by several orders of magnitude. The maximum differential voltage varies over 3000(V) as the conductivity changes by three orders of magnitude. Such a change is indeed possible as the conductivity varies depending on temperature and radiation does rate.

If we judge that ESD is unavoidable, however, we have to move to the next stage. We use a charging simulation code to estimate the number of ESD events expected in orbit and use the number in the ESD test plan. This can be done in the following steps [18]:

1. Obtain the ESD inception threshold in terms of the differential voltage by a laboratory experiment or other methods.

2. Determine and list the space plasma environments that a given spacecraft will encounter in the orbit.

3. Run the charging simulation for each case of the space plasma environment.

4. Identify the plasma environment cases that give a differential voltage exceeding the ESD inception threshold.

5. For each case identified in step 4, calculate the charging time necessary to reach the threshold value.

6. Derive the probability of occurrence of each case identified in step 4 and the expected total duration in orbit.

7. Divide the expected total duration obtained in the step 6 by the charging time obtained in the step 5 to obtain the expected number of ESD events for each case exceeding the ESD inception threshold.

8. Sum the numbers of ESD events to derive the total number.

More details can be found in [18]. In the first and fifth step, a proper database regarding the space plasma environment should be consulted. An example for a GEO database is the Los Alamos National Laboratory (LANL) satellite data that are available on the Internet (data available online at http://cdaweb.gsfc.nasa .gov/). An example for a PEO database is the Defense Meteorological Satellite Program (DMSP) satellite data. The auroral electron data are available at http://sd-www.jhuapl.edu/Aurora/index.html. The ionospheric low-temperature ion data are also available at http://cindispace.utdallas.edu/DMSP/.

A discussion of Fig. 5.2 illustrates how the charging time can be calculated in the fifth step. In this example, MUSCAT is used to calculate the charging of a GEO satellite. The simulation starts from the condition where both the satellite and coverglass potentials are zero. As the charging starts, the differential voltage gradually builds up. If the ESD inception threshold is 400 V, the threshold is reached in 40 s. Once the threshold is attained, an ESD occurs somewhere on the solar paddle, and the satellite body potential goes back to zero. By assuming that the differential charging is initialized by one ESD, the process of charging and ESD is repeated as long as the harmful plasma environment continues.

Table 5.1 illustrates how the number of ESD events in one year is derived. Using the result of statistical analysis in the fifth step, the expected duration of each plasma environment can be calculated for each local time zone in GEO. For each case, the charging analysis in the fourth step provides the time to reach the threshold. In the table "N/A" indicates that the differential charging does not reach the threshold value even at the steady state. The number of ESDs for each case is calculated by dividing the duration by the time to reach the threshold.

The third application of the charging simulation is to investigate an in-orbit anomaly. Even after careful simulation and testing, satellite anomalies can still occur in orbit. Because anomaly investigation is a very labor-intensive and time-consuming effort, we have to narrow down the possible cause of an anomaly as soon as possible. If the result of the charging simulation indicates that the possibility of charging at the time of the anomaly is minimal, we can disregard spacecraft charging as the anomaly cause and divert the resource to other possibilities. If the simulation result tells that the charging possibility was high, however, further detailed study involving laboratory experiments might become necessary.

Figure 5.6 shows the result of a NASCAP charging simulation carried out during the anomaly investigation of a GEO satellite series [19]. Although the exact plasma environment was not known, the simulation result revealed that the differential voltage could become high enough to trigger ESD on a solar cell, which was later found to be the source of the short circuit in the solar-array string circuit.

In [20] Cooke carried out a simulation by POLAR code to investigate the operational anomaly of the microwave imager onboard the DMSP F13 satellite. The anomaly was attributed to charging of ungrounded Teflon® MLI surfaces due to auroral electrons and subsequent electrostatic discharge [21]. The POLAR simulation confirmed that the insulator surface could indeed charge to several kilovolts negative with respect to the spacecraft body potential. Figure 5.7 shows an example of the simulation results. The wake produced by the solar paddle prevents ions from entering the top surface of the body while the energetic auroral electrons charge the surface negatively. This simulation demonstrated that severe differential charging is possible in polar orbit.

As simulation tools have become more sophisticated, charging simulation tools have been used for various types of anomaly investigations. In [22] NASCAP-2K was used to investigate an anomaly of the Special Sensor Ultraviolet Limb Imager (SSULI) onboard DMSP F16. It was suspected that ambient oxygen ions might have hit the detector when they were attracted by a negative chassis potential of the instrument when the spacecraft charged to a negative potential. NASCAP-2K was applied to calculate the trajectories of the ion particles and determined that the proposed mechanism was unlikely as the cause of the anomaly.

The fourth use of the simulation code is to interpret the scientific measurement data obtained by onboard instruments. The scientific measurement of the

**TABLE 5.1 EXAMPLE OF DERIVING THE NUMBER OF ESD EVENTS FOR A GEO SATELLITE (FROM TABLE 7 OF [18]. COPYRIGHT © 2005 AIAA. REPRINTED WITH PERMISSION.)**

| Plasma Parameters | | | | Duration in One Year, s | | | | | Time to Reach $\Delta V = 400$ V, s | | | | | Number of Arcs in One Year | | | | |
|---|---|---|---|---|---|---|---|---|---|---|---|---|---|---|---|---|---|---|
| $T_e$, keV | $n_e$, cm$^{-3}$ | $T_i$, keV | $n_i$, cm$^{-3}$ | LT0e | LT0n | LT6 | LT12 | LT18 | LT0e | LT0n | LT6 | LT12 | LT18 | LT0e | LT0n | LT6 | LT12 | LT18 |
| 13.5 | 5 | 5 | 0.25 | 0 | 290 | 106 | 439 | 33 | 8 | 9 | 10 | 9 | 9 | 0 | 32 | 11 | 49 | 4 |
| 10.5 | 10 | 10 | 0.25 | 0 | 0 | 448 | 156 | 29 | 5 | 6 | 6 | 6 | 6 | 0 | 0 | 75 | 26 | 5 |
| 10.5 | 5 | 10 | 0.25 | 4 | 55 | 558 | 507 | 194 | 9 | 12 | 12 | 12 | 12 | 0 | 5 | 47 | 42 | 16 |
| 10.5 | 5 | 5 | 0.25 | 0 | 154 | 321 | 405 | 70 | 9 | 12 | 12 | 12 | 12 | 0 | 13 | 27 | 34 | 6 |
| 7.5 | 10 | 10 | 0.25 | 4 | 13 | 220 | 329 | 70 | 7 | 8 | 8 | 8 | 8 | 1 | 2 | 28 | 41 | 9 |
| 7.5 | 10 | 5 | 0.25 | 0 | 34 | 313 | 173 | 83 | 7 | 8 | 8 | 8 | 8 | 0 | 4 | 39 | 22 | 10 |
| 7.5 | 5 | 10 | 0.25 | 8 | 111 | 1848 | 925 | 112 | 13 | 18 | 20 | 19 | 18 | 1 | 6 | 92 | 49 | 6 |
| 7.5 | 5 | 5 | 0.25 | 4 | 145 | 773 | 650 | 157 | 13 | 18 | 20 | 19 | 18 | 0 | 8 | 39 | 34 | 9 |
| 4.5 | 10 | 10 | 0.25 | 0 | 38 | 525 | 215 | 21 | 13 | 20 | 22 | 21 | 20 | 0 | 2 | 24 | 10 | 1 |
| 4.5 | 10 | 5 | 0.25 | 0 | 90 | 399 | 177 | 103 | 13 | 20 | 22 | 21 | 20 | 0 | 5 | 18 | 8 | 5 |
| 4.5 | 5 | 10 | 0.25 | 0 | 175 | 0 | 3184 | 673 | 26 | 140 | N/A | 170 | 140 | 0 | 1 | 0 | 19 | 5 |
| 4.5 | 5 | 5 | 0.25 | 4 | 249 | 0 | 887 | 549 | 26 | 120 | N/A | 140 | 120 | 0 | 2 | 0 | 6 | 5 |
| 4.5 | 5 | 1.25 | 0.25 | 0 | 148 | 0 | 0 | 0 | 26 | 110 | N/A | 130 | 110 | 0 | 1 | 0 | 0 | 0 |

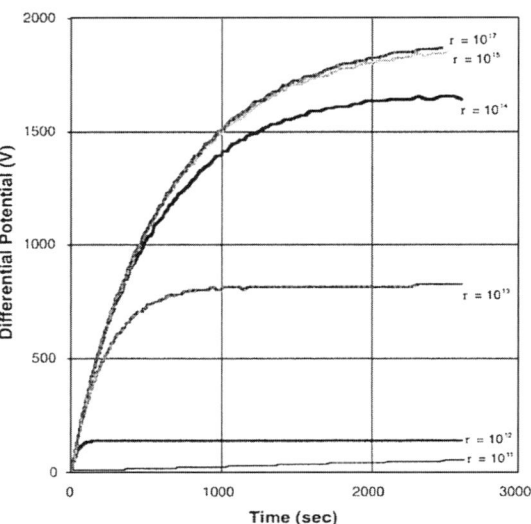

FIG. 5.6    Result of NASCAP-GEO simulation used to calculate the differential voltage during anomaly investigation of a GEO satellite series in 1997. (From Fig. 4 of [19]. Copyright © 1998 AIAA. Reprinted with permission.)

ambient space plasma environment is often hindered by the presence of the spacecraft itself. Even though significant care is taken to avoid differential charging on spacecraft by careful selection of surface materials or extensive use of conductive coatings, absolute charging of the spacecraft body cannot be avoided. The energy spectrum of a particle detector is shifted, and the electric field from the spacecraft body affects measurements of the weak electric fields in the ambient plasma. When a charging control device such as an ion emitter is used, the presence of the man-made charged particles must be accounted for in data interpretation. In [23], SPIS was applied to interpret the plasma measurement data of DEMETER. The influence of the metallic boom was analyzed.

## 5.5    NUMERICAL TECHNIQUES

The basis of the spacecraft charging simulation is to solve a set of differential

FIG. 5.7    Example of POLAR simulation carried out for the DMSP F13 satellite anomaly investigation. The surface potential distribution of the wake side view is shown. Note the highly negative surface potential behind the solar panel. (From Fig. 6 of [20]. Courtesy of D. Cooke.)

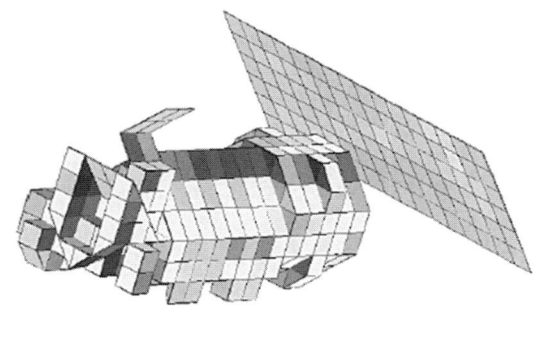

0.0        −5.0E+02      −1.0E+03        −1.5E+03       −2.0E+03      −2.5E+03

equations. To provide the solutions within a reasonable time frame, various assumptions and numerical techniques are employed. We will briefly introduce some of them.

The first step in the simulation is to define the spacecraft geometry and surface material properties. All of the modern charging simulation codes such as MUSCAT, NASCAP-2K, and SPIS have a GUI operable on a PC. Figure 5.8 shows an example of the GUI of MUSCAT. The users input data with a mouse. Because the GUI program is written in Java3D, it can run on any platform such as Windows®, Macintosh®, or Linux. NASCAP-2K and SPIS have a more or less similar capability. How easily the GUI can be used determines how widely the software will be used by people who are not familiar with spacecraft charging simulations. During code development, significant efforts have been spent in developing GUIs.

Once the spacecraft geometry is given, the geometry is mapped onto the numerical grid system. Figures 5.9, 5.10, and 5.11 show grid systems made for various satellites by NASCAP-2K, MUSCAT, and SPIS, respectively. The improvement from Polar to NASCAP is seen as we compare Figs. 5.9 and 5.8. Figure 5.10 is the grid system of the same satellite shown in Fig. 5.4, showing the grid system made by NASCAP/GEO. We can see the improvement made in the 30 years between NASCAP/GEO and MUSCAT. NASCAP-2K and SPIS employ a finite element method to solve the Poisson equation and derive the

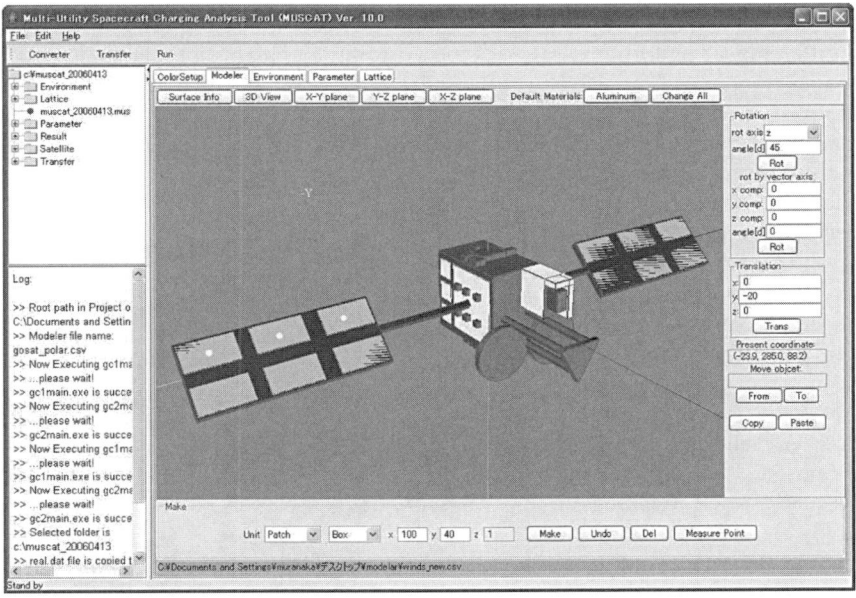

**FIG. 5.8   Snapshot of satellite modeling GUI front end of MUSCAT.**

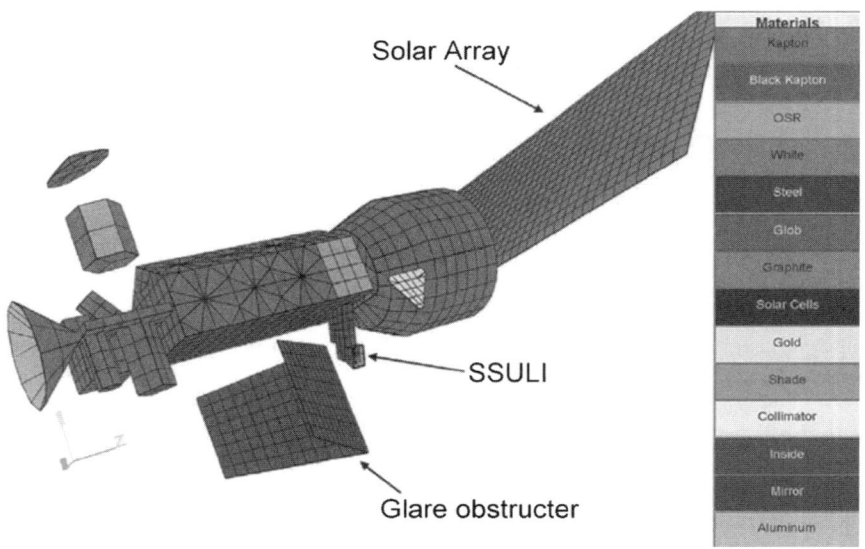

**FIG. 5.9    Example of numerical grid system made by NASCAP-2K for a DMSP satellite. (From Fig. 3 of [11]. © 2006 IEEE. Reprinted with permission and courtesy of SAIC.)**

electric potential. MUSCAT employs a finite difference method. This comes from a difference of philosophy. MUSCAT tried to find a balance between speed and spatial resolution. The use of a rectangular grid system has two advantages. The first one is that we can use fast Fourier transform for the Poisson solver, which eliminates costly numerical iterations. The use of a finite element method inherently requires numerical iteration. The second advantage is that following the charged particle trajectory becomes much simpler and faster compared to an adaptive grid system, where even finding the grid index numbers the particle should refer to interpolate the local electric field is not a simple task. A technique has been proposed to combine the ease of particle tracking in

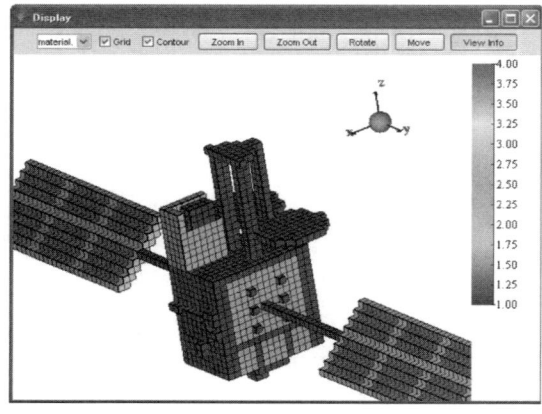

**FIG. 5.10    Example of numerical grid system made by MUSCAT for WINDS.**

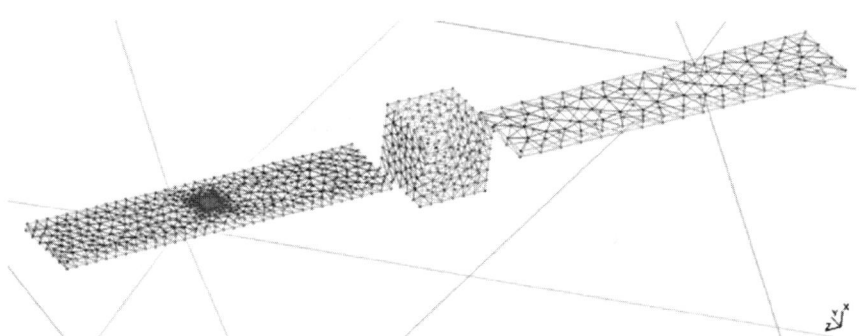

**FIG. 5.11** **Example of numerical grid system made by SPIS. Note the local refinement around a gap of a solar array. (From Fig. 10 of [14]. © 2008 IEEE. Reprinted with permission.)**

a rectangular mesh system and the fineness of the potential structure in the adaptive mesh system [24]. This technique can be adopted in the next generation of charging simulation codes.

When we integrate the differential equations [Eqs. (5.2)], we need to define each current element to the surface. The particle-in-cell (PIC) method [25, 26] has been used extensively to follow the motion of macroparticles that represent a group of particles chosen from the velocity distribution function. PIC methods put some constraints on the time step and the grid size. The time step must be less than a fraction of the inverse of the electron plasma frequency, and the numerical grid size must be less than the Debye length. Solving the differential equations (5.2) by a PIC code alone is not yet possible as the time step defined by the electron plasma frequency is still too small to cover the entire charging profile such as the one shown in Fig. 5.2. To cope with this difficulty, for example, MUSCAT sacrifices accuracy within the inverse of the ion plasma frequency. Once the PIC simulation defines the sheath boundary structure, simple particle tracking is carried out to calculate the current to each surface element. In the particle tracking, the spatial potential is fixed, which means that the sheath structure does not vary significantly during the timescale for ions to travel from the sheath boundary to the spacecraft surface. In particle tracking we can use a much longer time step to integrate the equation of motion than in PIC simulation [3]. Therefore, there can be three time steps in 1) the PIC simulation, 2) the particle tracking, and 3) the numerical integration of the differential equations (5.2). In this way we can integrate Eqs. (5.2) free from the restrictions of particle dynamics.

When the plasma is sufficiently tenuous as in GEO, we no longer need to consider space charge effects. In that case, we don't need a PIC method to calculate the surface current, thus greatly reducing the simulation time. In a tenuous plasma, the field equation also becomes the Laplace equation instead of the Poisson equation. NASCAP-2K uses the boundary element method for the tenuous plasma to save the computation time needed to define the potential structure.

In particle tracking, reverse tracking is often mentioned as a technique of further speed-up [13]. Reverse tracking was employed by NASCAP/GEO [27]. When we follow the particle trajectories from the outer boundary, many particles miss the satellite surface and waste computational time. In reverse particle tracking we integrate the equation of motion backward from the spacecraft surface to the outer boundary. The distribution function at the outer boundary can be well approximated by a known distribution such as a Maxwellian or a double Maxwellian. If we know the distribution function at the spacecraft surface and can match the volume elements of the two distribution functions at both of the boundaries, we can derive the current to the surface by reverse tracking. However, the particle distribution function at the spacecraft surface is not known a priori for a complex spacecraft. Therefore, one must be careful about the error associated with the assumption of the distribution function at the spacecraft surface.

MUSCAT employs two other acceleration techniques, that is, parallelization and a variable time step. MUSCAT is developed for a general-purpose symmetric multiple processor (SMP) computer, and the solver is parallelized on the OpenMP specification. There are three main parts that benefit from the parallelization. They are the PIC method, the particle-tracking (PT) method, and the field solver. MUSCAT parallelizes the particle movers based on the domain decomposition scheme. Because an OpenMP command is specified as special comment lines in a Fortran source code, MUSCAT can also run on a single-processor machine, or even on a laptop computer, by recompiling the source code with the parallelization option set to "off." MUSCAT varies the time step for integration of the differential equations (5.2) while ensuring the stable development of charging potentials.

## 5.6  VALIDATION

Any computer simulation code used for engineering design must demonstrate its accuracy through careful validation processes using either experiments or theories or both. The most fundamental validation process is comparison with Langmuir probe theory or experiment. Figures 5.12–5.14 show the comparison with theory and experiment for a simple probe. All of the results show good agreement with the references, though this is only the first step. It is best to use a laboratory test result to validate the accuracy of the physical model and numerical schemes used in each simulation code rather than comparing the simulation with on-orbit data because it is often difficult to identify all of the relevant input parameters for the simulation from the on-orbit measurement data. We can modify the laboratory experiment so that the situation exactly matches the input parameters of the simulation. Figure 5.15 shows such an example where a rectangular plate of $44 \times 44$ mm was placed in a dense flowing plasma of the order of $10^{15}$ m$^{-3}$,

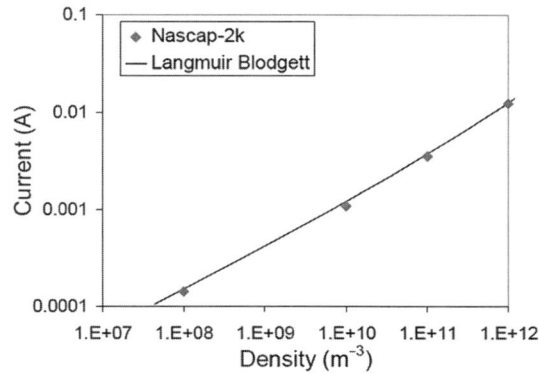

FIG. 5.12  Comparison of NASCAP-2K with Langmuir–Blodgett theory. (From Fig. 1 of [12]. Courtesy of SAIC.)

simulating a solar panel of 7.6×7.6 m from a scaling law [28]. A wake structure was produced behind the plate, and the simulation result matches the experiment very well.

The accuracy of numerical integration of the differential equations (5.2) can be partially validated by a simple spherical model in the orbital-motion-limited (OML) regime. Let us suppose a sphere with radius $R$. The capacitance of the sphere with respect to infinity is $4\pi R/\varepsilon_o$. Considering only the ambient electrons and ions, the temporal profile of the sphere potential is given by

$$\frac{\mathrm{d}\phi_s}{\mathrm{d}t} = \frac{R}{\varepsilon_o}\left[j_{0e}\exp\left(\frac{e\phi_s}{kT_e}\right) - j_{0i}\left(\frac{1-e\phi_s}{kT_i}\right)\right] \tag{5.5}$$

where the sphere potential $\phi_s$ is negative. The current densities $j_{0e}$ and $j_{0i}$ are the current density in the far field and given by

$$j_{oe,i} = \frac{1}{\sqrt{2\pi}}en_{e,i}\sqrt{kT_{e,i}/m_{e,i}} \tag{5.6}$$

assuming a Maxwellian distribution. Figure 5.16 shows a comparison between the potential given by OML theory and the numerical result calculated by MUSCAT, where a conductive sphere of 1-m radius was placed in a tenuous plasma of $1.25\times10^6$ m$^{-3}$ density. The electron and ion temperatures were 7.5 and 10 keV, respectively.

Although laboratory experiments are best to validate the accuracy of the

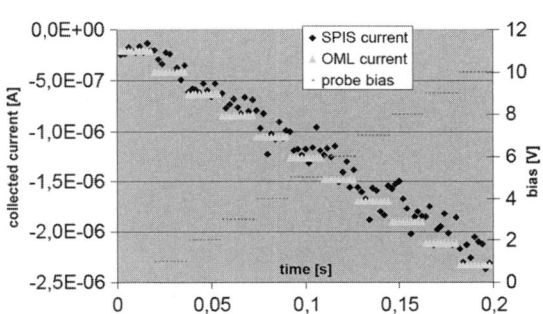

FIG. 5.13  Comparison of SPIS with orbital motion limited theory. (From Fig. 4 of [13]. Courtesy of J. F. Roussel.)

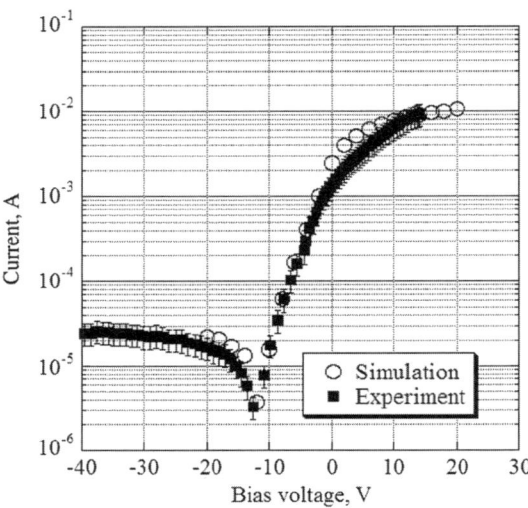

FIG. 5.14 Comparison of MUSCAT with experiment. (From Fig. 10 of [28]. © 2008 IEEE. Reprinted with permission.)

numerical solver, the final validation should be done against on-orbit data. This is not because of accuracy, but to confirm that the simulation code can explain the on-orbit phenomena without overlooking any important physical processes that might affect the charging processes. Various comparisons between NASCAP/GEO simulations and data from the SCATHA satellite [29] have been made, which contributes to the wide use of NASCAP/GEO throughout the world.

To validate a charging simulation code by on-orbit data, we at least need instruments to measure the spacecraft chassis potential and the plasma parameters. (Density and temperature are the minimum requirement.) If a satellite has an instrument to measure the particle energy spectra, the chassis potential can be measured from the shift in the peak particle energies, though the energy range should cover below 30 keV and the resolution must be sufficiently high, hopefully 10 V or less. An instrument to measure the insulator charging potential [30] is also desirable to validate the capability to predict differential charging. The telemetry interval must be short enough to resolve the development of the differential charging buildup as shown in Fig. 5.2, 1 s or less, if we want to validate the temporal accuracy of the simulation code. Information regarding the surface material (including thickness if it is insulator) and geometry must be provided to formulate a proper simulation model.

FIG. 5.15 Comparison of the space potential behind a plate simulating a solar panel in PEO as measured in a laboratory experiment and as found in a MUSCAT simulation. (From Fig. 18 of [28]. © 2008 IEEE. Reprinted with permission.)

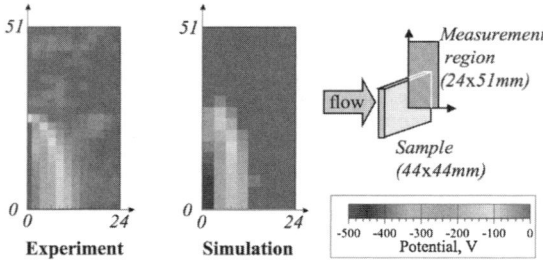

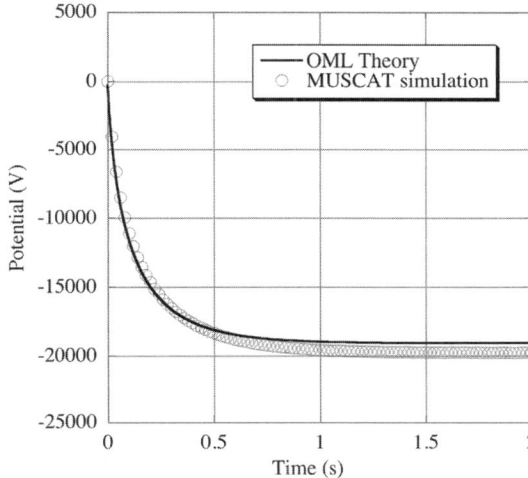

FIG. 5.16  Temporal profile of a spherical body potential calculated by MUSCAT and OML theory. (From Fig. 9 of [4]. © 2008 IEEE. Reprinted with permission.)

## 5.7  KEY POINTS TO A GOOD CHARGING SIMULATION CODE AND FURTHER IMPROVEMENTS

There are several key points to a good charging simulation code. The first one is user friendliness. For any computer simulation code to aid engineering design, the users are usually neither experts on the physical processes they are dealing with nor experts in computer technology. The code must be easy to install on the user's PC without any special assistance from the vendor or computer experts. The input parameters must be easy to understand, and no special knowledge about the computer language should be necessary to specify the input parameters.

A second key point is graphical capability in spacecraft modeling and data visualization. The graphical response must be fast enough not to irritate the users. One desirable point is the capability to import a CAD data file. All of MUSCAT, NASCAP-2K, and SPIS ask the users to formulate the satellite model by combining basic components such as boxes, cylinders, and other simple geometrical shapes. Although the task would not take more than a couple of hours, if the users become familiarized with the procedures, users always try to find an easier path. Because every spacecraft is designed based on CAD data files, users often ask about the possibility of CAD import functions. Because CAD data files usually contain much more data than needed for the charging simulation, a capability of importing CAD data and function to select only necessary information would attract more users to charging simulations.

Of course, having the functionality of three-dimensional visualization of the surface potential and other properties is highly desirable. In addition, the simulation code should be equipped with an interactive data visualization tool that enables users to examine the data closely by selecting a point or element on the spacecraft surface to examine the temporal profile of charging history or each current component to the surface element. An example of MUSCAT data visualization is shown in Fig. 5.17. In this example, the temporal profile of the body potential and the surface potential of a point that is specified in advance is shown. We can also plot the temporal profiles of conduction current, indicated by condcrnt, low-energy part electron current, indicated by electron(l), photo-electron current, indicated by photo, and so on, from a menu.

A third key point is the material database. Even if the code is equipped with a state-of-the-art numerical solver, the simulation result will still be incorrect if the material properties are incorrect. The most important material properties are secondary electron coefficient, photoelectron coefficient, and bulk conductivity. Figure 5.18 shows the material properties data window of MUSCAT. This window first pops up when the user starts the MUSCAT GUI program. The window already contains data for some of the commonly used materials collected from the literature. But it is far from perfect. Currently there are campaigns to measure the key material property data in several countries [31, 32]. Exchange of material-properties databases is desirable to avoid duplication of effort. One important point to be noted is that all of the measurements currently have been carried out on virgin samples. In reality, every spacecraft surface material degrades as a result of long time exposure to the space environment. If we plan to extend the charging simulation to the situation at the end of life in orbit, a database of aged materials becomes necessary. Such an effort should be started very soon.

As Fig. 5.5 indicates, the conductivity of coverglasses makes a big difference in a GEO charging simulation. It has been well known that insulator conductivity increases under radiation exposure [33, 34] as electrons are excited to the conduction band by absorbing the radiation energy. The radiation-induced conductivity (RIC) at GEO can greatly affect the result of a charging simulation. RIC is often modeled by the following equation:

$$\sigma = \sigma_o + k_p D^\Delta \tag{5.7}$$

where $\sigma_o$ is the dark conductivity where no radiation effect exists, $k_p$ is a coefficient dependent on each material, $D$ is the radiation dose rate, and $\Delta$ is a nondimensional factor less than unity dependent on each material. Most of the available data for RIC were measured under a radiation dose much higher than the

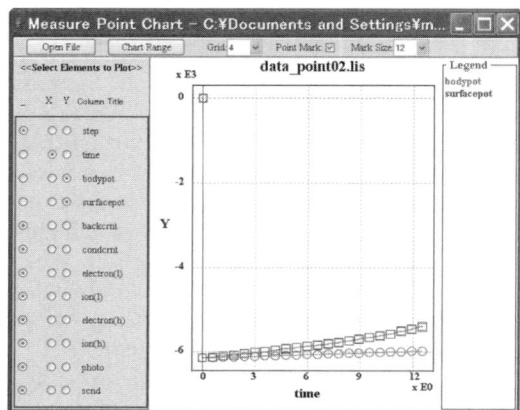

**FIG. 5.17  Example of MUSCAT data visualization screen. Temporal profiles of various quantities such as potential and currents of each component at prespecified points on spacecraft surface can be plotted.**

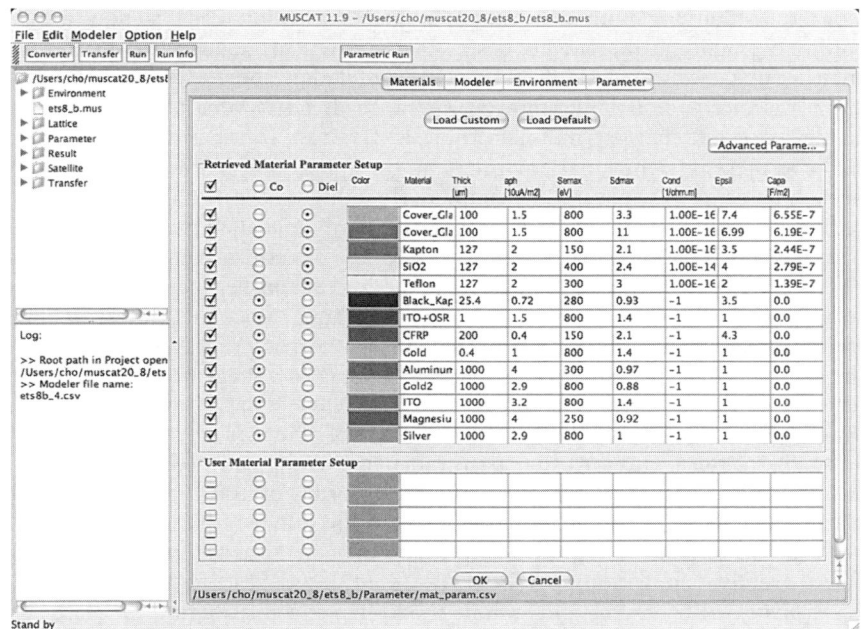

**FIG. 5.18    Material selection window of MUSCAT.**

dose rate in orbit. We need more reliable data on RIC for spacecraft surface dielectrics.

The fourth key point is the environmental database. MUSCAT, NASCAP-2K, and SPIS specify a given set of plasma environment parameters for each run. The users are supposed to specify the plasma environment with some helps from the manual or the online help menu. When we carry out parametric runs for many combinations of the plasma parameters, it is too much burden on the user to specify each environment set. MUSCAT has CSV files that contain possible combinations of densities and temperatures of the plasma for GEO and PEO [18, 35]. Once a user writes a script file in a UNIX shell language, MUSCAT can run all of the cases automatically. The next step would be one-click capability. It would be very nice if software automatically calculated all of the possible cases and gave the answer about the maximum charging potential, the total duration of charging event, the expected number of ESD events, etc., once a user clicked a button of "parametric runs."

## 5.8   FUTURE SUBJECTS

Spacecraft charging has been successfully applied at various stages of spacecraft development and operation. Especially for GEO satellites, calculating the worst-case charging potential via a charging simulation code is now a very common practice at the satellite design phase. There are many other interesting subjects that await analysis by a charging simulation code. The first subject is application to the lunar environment. There are already several publications analyzing spacecraft charging in lunar orbit [36, 37]. If we consider only the charged particles present in the lunar environment, the application is straightforward: we need only identify the plasma environment (that might be little different from that of Earth orbit). Once we try to study the effects of lunar dust particles, the work becomes much harder as we have to consider the dynamics of dusty plasmas.

A second interesting future subject is the interaction between a high-voltage solar array and an electric propulsion thruster plume. Since the success of Deep Space-1, there have been many concepts of solar electric propulsion. To increase the system efficiency, a high-voltage power generation voltage over 100 V is preferred, and even a concept of direct-drive by a 300-V solar array has been proposed [38]. The charge-exchange ions in the thruster plume diffuse via ambipolar diffusion toward the spacecraft body. The positive end of the solar-array circuit can expose a high voltage of the order of 100 V to the diffusing plasma and strongly interact with electrons in the diffusing plasma. There have been many simulations of the electric-propulsion thruster plume [39–41]. There are also simulations by NASCAP-2K [42] and SPIS [6]. As long as we use a Hybrid-PIC scheme and assume the Boltzmann distribution for electrons, we cannot resolve the issue of interaction between the neutralizer electrons and the solar array. Either we solve the problem by a full-PIC scheme or we need a new idea to solve the problem with affordable computer resources. There are also many modeling efforts for EP plume itself [43, 44]. MUSCAT, NASCAP-2K, and SPIS all have an option to calculate the thruster plume interaction. Some day in the near future, there might be a simulation code that can cover everything from the plasma production in the discharge chamber of the electric propulsion system to the backflow flux to the spacecraft exterior surface.

The third future subject is to simulate how the surrounding plasma reacts to an abrupt change of spacecraft potential due to ESD or other unpredicted reasons to assess their EMC effects. The timescale involved here is as short as microseconds or less. Therefore, resolving the motion of electrons self-consistently with its space charge becomes necessary. As this task requires a tremendous amount of computer memory, it has not been fulfilled yet at a practical level, although some efforts are underway using a massive parallel computer [16].

The fourth future subject is cross-validation of different charging simulation codes. Until recently, there was only NASCAP/GEO available for everybody in the world. Therefore, there was no problem about what code we chose. Nowadays, the charging simulation codes have been renewed to MUSCAT, NASCAP-2K, SPIS, and others worldwide. In the very near future, satellite design guidelines of various countries will require assessing the risk of ESD in orbit using a proper charging simulation code. Then it will only be a matter of time to specify what simulation code should be used in a contract involving more than two countries. Cross-validation among the codes would become necessary if all of the codes are to survive. The effort to coordinate the cross-validation started in 2005 at the 9th Spacecraft Charging Technology Conference but has made little progress. Motivation will be renewed soon, as the next-generation charging codes enter the stage of practical use and the commercialization of the software progresses.

## 5.9  CONCLUSION

The basic requirement of spacecraft charging simulations is to find the spacecraft potential with respect to the surrounding plasma. In addition, there are needs to find the potentials of surface insulators and the near-spacecraft plasma conditions. Obtaining the temporal profiles of those parameters is more desirable than obtaining final state solutions only. Various spacecraft charging simulation tools have been developed exceeding the capability of NASCAP/GEO that served as the de facto standard for nearly 20 years, employing more user-friendly graphic user interface, better and faster algorithm, and finer spatial resolution. Charging simulation involves knowledge of plasma physics, particle-surface interactions, and computer algorithms. A good simulation tool should be, however, capable of providing a correct answer in a short time to a user without deep knowledge in the just-mentioned fields. For the answer to be trusted, every simulation code has to be thoroughly validated before practical use. The material database has the most significant effect on the accuracy of the simulation result. Therefore, constant efforts to validate, modify, and expand the database are as important as efforts to expand the computational capability and improve the numerical techniques and user interface.

## REFERENCES

[1]  Langmuir, I., and Blodgett, K. B., "Current Limited by Space Charge Flow Between Concentric Spheres," *Physical Review*, Vol. 24, No. 1, 1924, pp. 49–59.

[2]  Laframboise, J. G., and Sonmar, L. J., "Current Collection by Probes and Electrodes in Space Magnetoplasmas: A Review," *Journal of Geophysical Research*, Vol. 98, No. A1, 1993, pp. 337–357.

[3]  Cho, M., and Hastings, D. E., "Dielectric Charging Processes and Arcing Rates of High Voltage Solar Arrays," *Journal of Spacecraft and Rockets*, Vol. 28, No. 6, 1991, pp. 698–706.

[4]  Muranaka, T., Hosoda, S., Kim, J., Hatta, S., Ikeda, K., Hamanaga, T., Cho, M., Usui, H., Ueda, O. H., Koga, K., and Goka, T., "Development of Multi-Utility Spacecraft Charging Analysis Tool (MUSCAT)," *IEEE Transactions on Plasma Science*, Vol. 36, No. 5, 2008, pp. 2336–2349.

[5]  Cho, M., Ramasamy, R., Matusmoto, T., Toyoda, K., Nozaki, Y., and Takahashi, M., "Laboratory Tests on 110 V Solar Arrays in a Simulated Geosynchronous

Orbit Environment," *Journal of Spacecraft and Rockets*, Vol. 40, No. 2, 2003, pp. 211–220.

[6]  Hilgers, A., Thiebault, B., Estublier, D., Gengembre, E., Gonzalez, J., Tajmar, M., Roussel, J.-F., and Forest, J., "A Simple Model of SMART-1 Electrostatic Potential Variation," *IEEE Transactions on Plasma Science*, Vol. 34, No. 5, 2006, pp. 2159–2165.

[7]  Besse, A. L., and Rubin, A. G., "A Simple Analysis of Spacecraft Charging Involving Blocked Photoelectron Currents," *Journal of Geophysical Research*, Vol. 85, No. A5, 1980, pp. 2324–2328.

[8]  Katz, I., Cassidy, J. J., Mandell, M. J., Schnuelle, G. W., Steen, P. G., and Roche, J. C., "The Capabilities of the NASA Charging Analyzer Program," *Spacecraft Charging Technology-1978*, edited by R. C. Finke and C. P. Pike, Air Force Geophysics Lab., Hanscom AFB, MA, 1979, p. 101; also NASA CP-2071/AFGL TR-79-0082, ADA045459.

[9]  Katz, I., Mandell, M. J., Schnuelle, G. W., Parks, D. E., and and Steen, P. G., "Plasma Collection by High-Voltage Spacecraft at Low Earth Orbit," *Journal of Spacecraft and Rockets*, Vol. 18, No. 1, 1981, pp. 79–82.

[10]  Lilley, J. R., Cooke, D. L., Jongeward, G. A., and Katz, I., "POLAR User's Manual," Air Force Geophysics Lab., AFGL-TR-85-0246, Hanscom AFB, MA, Oct. 1985.

[11]  Davis, V. A., Mandell, M. J., Rich, F. J., and Cooke, D. L., "Reverse Trajectory Approach to Computing Ionospheric Currents to the Special Sensor Ultraviolet Limb Imager on DMSP," *IEEE Transactions on Plasma Science*, Vol. 34, No. 5, 2006, pp. 2062–2070.

[12]  Davis, V. A., Mandell, M. J., Gardner, B. M., Mikellides, I. G., Neergaard, L. F., Cooke, D. L., and Minor, J., "Validation of NASCAP-2K Spacecraft-Environment Interactions Calculations," *8th Spacecraft Charging Technology Conference* [CD-ROM], edited by J. Minor, NASA Marshall Space Flight Center, Huntsville, AL, 2003; also NASA/CP—2004–213091.

[13]  Roussel, J-F., Rogier, F., Volpert, D., Forest, J., Rousseau, G., and Hilgers, A., "Spacecraft Plasma Interaction Software (Spis): Numerical Solvers–Methods and Architecture," *9th Spacecraft Charging Technology Conference*, edited by T. Goka, Vol. 1, Japan Aerospace Exploration Agency, Tsukuba, Japan, 2005, pp. 462–472; also JAXA-SP-05-001E.

[14]  Roussel, J-F., Rogier, F., Dufour, G., Mateo-Velez, J-C., Forest, J., Hilgers, A., Rodgers, D., Girard, L., and Payan, D., "SPIS Open-Source Code: Methods, Capabilities, Achievements, and Prospects," *IEEE Transactions on Plasma Science*, Vol. 36, No. 5, 2008, pp. 2360–2368.

[15]  Clerc, S., Brosse, S., and Chane-Yook, M., "SPARCS: An Advanced Software for Spacecraft Charging Analyses," *8th Spacecraft Charging Technology Conference* [CD-ROM], edited by J. Minor, NASA Marshall Space Flight Center, Huntsville, AL, 2003; also NASA/CP—2004–213091.

[16]  Usui, H., Miyake, Y., Okada, M., Omura, Y., Sugiyama, T., Murata, K. T., Matsuoka, D., and Ueda, H. O., "Development and Application of Geospace Environment Simulator for the Analysis of Spacecraft–Plasma Interactions," *IEEE Transactions on Plasma Science*, Vol. 34, No. 5, 2006, pp. 2094–2102.

[17]  Purvis, C. K., Garret, H. B., and Whittlesey, A. C., "Design Guidelines for Assessing and Controlling Spacecraft Charging," NASA-TP-2361, Sept. 1984.

[18]    Cho, M., Kawakita, S., Nakamura, S., Takahashi, M., Sato, T., and Nozaki, Y.,
        "Number of Arcs Estimated on Solar Array of a Geostationary Satellite," *Journal of
        Spacecraft and Rockets*, Vol. 42 No. 4, 2005, pp. 740–748.

[19]    Hoeber, C. F., Robertson, E. A., Katz, I., Davis, V. A., and Snyder, D. B., "Solar Array
        Augmented Electrostatic Discharge in GEO," AIAA Paper 98-1401, Feb. 1998.

[20]    Cooke, D. L., "Simulation of an Auroral Charging Anomaly on the DMSP Satellite,"
        *6th Spacecraft Charging Technology Conference Proceedings*, edited by D. L. Cooke,
        AFRL-VS-TR-20001578, U.S. Air Force Research Lab., Hanscom AFB, MA, 1998,
        pp. 33–37.

[21]    Anderson, P. C., and Koons, H. C., "Spacecraft Charging Anomaly on a
        Low-Altitude Satellite in an Aurora," *Journal of Spacecraft and Rockets*, Vol. 33, No.
        5, 1996, pp. 734–738.

[22]    Davis, V. A., Mandell, M. J., Rich, F. J., and Cooke, D. L., "Reverse Trajectory
        Approach to Computing Ionospheric Currents to the Special Sensor Ultraviolet
        Limb Imager on DMSP," *9th Spacecraft Charging Technology Conference*, edited by
        T. Goka, Vol. 1, Japan Aerospace Exploration Agency, Tsukuba, Japan, 2005,
        pp. 890–903.

[23]    Berthelier, J. J., Forest, J., Chen, X., Yang, J., and Quassim, M., "Demeter
        Measurements in Equatorial Plasma Depletions, Numerical Simulation of the
        Electric Field Probe Operation," 10th Spacecraft Charging Technology Conference,
        Paper 76, June 2007.

[24]    Kafafy, R., and Wang, J., "A Hybrid Grid Immersed Finite Element Particle-in-Cell
        Algorithm for Modeling Spacecraft–Plasma Interactions," *IEEE Transactions on
        Plasma Science*, Vol. 34, No. 5, Part 2, 2006, pp. 2114–2124.

[25]    Birdsall, C. K., and Langdon, A. B., *Plasma Physics via Computer Simulation*,
        McGraw–Hill, New York, 1985, p. 21.

[26]    Hockney, R. W., and Eastwood, J. W., *Computer Simulation Using Particles*, Adam
        Hilger, New York, 1988, p. 125.

[27]    Rubin, A. G., Katz, I., Mandell, M., Schunelle, G., Steen, P., Parks, D., Cassidy, J., and
        Roche, J., "A Three-Dimensional Spacecraft-Charging Computer Code," *Space
        Systems and Their Interactions with Earth's Space Environment*, edited by
        H. B. Garret and C. P. Pike, Progress in Astronautics and Aeronautics, Vol. 71,
        AIAA, New York, 1980, pp. 318–336.

[28]    Hosoda, S., Muranaka, T., Kim, J., Hatta, S., Kurahara, N., Cho, M., Ueda, H. O.,
        Koga, K., Goka, T., and Kuninaka, H., "Laboratory Experiments for Code Validation
        of Multi-Utility Spacecraft Charging Analysis Tool (MUSCAT)," *IEEE Transactions
        on Plasma Science*, Vol. 36, Oct. 2008, pp. 2350–2359.

[29]    Stannard, P. R., Katz, I., Gedeon, L., Roche, J. C., Rubin, A. G., and Tautz, M. F.,
        "Validation of the NASCAP Model Using Spaceflight Data," AIAA Paper 82-0269,
        Jan. 1982.

[30]    Bogorad, A., Bowman, C., Dennis, A., Beck, J., Lang, D., Herschitz, R., Buehler, M.,
        Blaes, B., and Martin, D., "Integrated Environmental Monitoring System for
        Spacecraft," *IEEE Transactions on Nuclear Science*, Vol. 42, No. 6, 1995, pp. 2051
        –2057.

[31]    Dennison, J. R., Thomson, C. D., Kite, J., Zavyalov, V., and Corbridge, J., "Materials
        Characterization at Utah State University: Facilities and Knowledgebase of
        Electronic Properties of Materials Applicable to Spacecraft Charging," *8th Spacecraft*

*Charging Technology Conference* [CD-ROM], edited by J. Minor, NASA Marshall Space Flight Center, Huntsville, AL, 2003; also NASA/CP—2004–213091.

[32] Miyake, H., Nitta, K., Michizono, S., and Saito, Y., "Secondary Electron Emission Measurement of Insulating Materials for Spacecraft," 10th Spacecraft Charging Technology Conference, Paper 24, June 2007.

[33] Fowler, J. F., "Analytic Expression for Electrons Transmission in Dielectrics," *Proceedings of the Royal Society of London*, Vol. A236, No. 1207, Sept. 1956, pp. 464–475.

[34] Weaver, L., Shultis, J., and Faw, R., "Analytic Solutions for Radiation Induced Conductivity in Insulators," *Journal of Applied Physics*, Vol. 48, No. 7, 1977, pp. 2762–2770.

[35] Hamanaga, T., and Cho, M., "Correlation Analysis of Energetic Electrons and Low Temperature Ions in Polar Orbit Using the Satellite Observation Data," 10th Spacecraft Charging Technology Conference, Paper 35, June 2007.

[36] Parker, L. N., Minow, J. I., and Blackwell, W. C., "Analysis of Lunar Surface Charging for a Candidate Spacecraft Using Nascap-2k," 10th Spacecraft Charging Technology Conference, Paper 311, June 2007.

[37] Wang, J., He, X., and Cao, Y., "Modeling Spacecraft Charging and Charged Dust Particle Interactions on Lunar Surface," 10th Spacecraft Charging Technology Conference, Paper 54, June 2007.

[38] Schneider, T. A., Mikellides, I. G., Jongeward, G. A., Peterson, T. P., Kerslake, T. W., Snyder, D., and Ferguson, D., "Solar Arrays for Direct-Drive Electric Propulsion: Arcing at High Voltages," *Journal of Spacecraft and Rockets*, Vol. 42, No. 3, 2005, pp. 543–549.

[39] Roy, R. I. S., Hastings, D. E., and Gastonis, N. A., "Ion-Thruster Plume Modeling for Backflow Contamination," *Journal of Spacecraft and Rockets*, Vol. 33, No. 4, 1996, pp. 525–534.

[40] Wang, J., Brinza, D. E., Young, D. T., Nordholt, J. E., Polk, J. E., Henry, M. D., Goldstein, R., Hanley, J. J., Lawrence, D. J., and Shappirio, M., "Deep Space One Investigations of Ion Propulsion Plasma Environment," *Journal of Spacecraft and Rockets*, Vol. 37, No. 5, 2000, pp. 545–555.

[41] Wang, J., Brinza, D., and Young, M., "Three-Dimensional Particle Simulations of Ion Propulsion Plasma Environment for Deep Space 1," *Journal of Spacecraft and Rockets*, Vol. 38, No. 3, 2001, pp. 433–440.

[42] Mandell, M. J., Davis, V. A., Pencil, E. J., Patterson, M. J., Mcewen, H. K., Foster, J. E., and Snyder, J. S., "Modeling the NEXT Multi-Thruster Array Test with Nascap-2K," *IEEE Transactions on Plasma Science*, Vol. 36, No. 5, 2008, pp. 2309–2318.

[43] Katz, I., Mikellides, I. G., Wirz, R., Anderson, J. R., and Goebel, D. M., "Ion Thruster Life Models," AIAA Paper 2005-4256, July 2005.

[44] Wang, J., Polk, J., Brophy, J., and Katz, I., "Three-Dimensional Particle Simulations of Ion-Optics Plasma Flow and Grid Erosion," *Journal of Propulsion and Power*, Vol. 19, No. 6, 2003, pp. 1192–1199.

# CHAPTER 6

# Spacecraft Charging in the Auroral Oval

Lars Eliasson[*]
*Swedish Institute of Space Physics, Kiruna, Sweden*

Anders I. Eriksson[†]
*Swedish Institute of Space Physics, Uppsala, Sweden*

## 6.1 INTRODUCTION

Several regions in the magnetosphere where spacecraft charging occurs have been identified. One example is the high-latitude magnetic field lines with energetic auroral particles that are traversed by satellites in polar Earth orbit (PEO). Charging effects in this region depend on several parameters, for example, the geomagnetic activity level, which affects the acceleration of auroral particle populations, and the solar illumination conditions because solar ultraviolet (UV) is the main ionization agent in the ionosphere. The high charging levels detected in geosynchronous orbit (GEO) were originally not expected to occur at low altitudes, where negative charging should be inhibited by the ion current from the dense ionosphere. However, in the auroral zone, such charging has been found to occur quite commonly.

In this chapter, we will summarize spacecraft charging processes in the auroral region. In Sec. 6.2, we first give an overview of the plasma environment and give the basics of why charging in fact does occur also in this region. After a brief Sec. 6.3, on the observed effects on satellites, we go deeper into the results of charging studies in the MEO (middle Earth orbit, between LEO and GEO) and LEO (low Earth orbit) altitude ranges in Secs. 6.4 and 6.5, based on the Freja and DMSP satellites, respectively. We summarize the main conclusions and discuss impacts on future mission and instrument design in Sec. 6.6.

## 6.2 ENVIRONMENT AND CHARGING IN THE AURORAL REGION

The spacecraft environment in the auroral region includes secondary electrons emitted from the surfaces, as well as the ambient thermal and energetic magnetospheric plasma, the plasma released from plasma thrusters, that are created by

---

[*]Director and Member of Technical Staff, P.O. Box 812, SE-981 28.
[†]Member of Scientific Staff, P.O. Box 537, SE-751 21.

ionization of or charge exchange with the expelled or ambient neutral gas, that are generated by arc discharges, and that are created by micrometeorite impacts. The solar UV radiation will also contribute to the plasma through emission of photo-electrons from the spacecraft surface. All of these sources affect the charge balance of the spacecraft.

Theoretical predictions of negative charging amounting to several kilovolts in polar LEO [1] were believed to present hazards for orbiting satellites and for manned polar flights [2]. It was nevertheless a surprise when data from the Defense Meteorological Satellites Program (DMSP) found many events above 50-V negative potential and even reaching values greater than a kilovolt negative within the auroral region at an altitude of only 840 km [e.g., 3–7].

The current flows to and from a spacecraft in different space environments can cause charge accumulation on the spacecraft. This charging process can produce potential differences between electrically isolated surfaces or between the spacecraft ground and the surrounding plasma. In extreme cases, the subsequent discharge can lead to total loss of a spacecraft. A recent and well-studied example is the loss of the ADEOS-II satellite in 2003, attributed to arc discharges following differential charging at 800-km altitude in the auroral zone [8, 9]. Less drastic but more common are problems for the operational abilities of the complex and sensitive payloads carried by modern scientific satellites, which can be affected by the space environment. Many instruments have experienced operation problems as a result of the plasma electrostatic environment, and the interpretation of data from measurements depends on knowledge of spacecraft charging effects. Any shift in potential relative to the spacecraft ground or the space plasma can affect instruments designed to collect or emit charged particles. The spacecraft environment can also be a source of instrumental noise in general. Besides destructive arcing effects on surfaces, a charge buildup on a spacecraft can in itself attract charged contaminants to sensitive surfaces that can alter the properties of the surface.

The spectral characteristics of a typical auroral active event often have insignificant low-energy fluxes (<1 keV) and large fluxes of high-energy electrons (up to a few tens of kiloelectron volts, Fig. 6.1). These properties are effective in producing high charging levels on spacecraft where the secondary yield crossover energy for the surface material is only a few kiloelectron volts. Highly conductive and grounded surface materials are needed to avoid differential charging problems between different surfaces.

Several factors contribute to producing conditions favorable for causing spacecraft charging at LEO and MEO altitudes in the auroral zone. The plasma charge shielding distance (Debye length) in the ionosphere is of the order of centimeters, as compared to meters or tens of meters at geostationary altitudes. The effective current collecting area of a spacecraft in LEO will thus be much smaller than in GEO. In addition, satellites in LEO orbit move supersonically with respect to the plasma ions, forming a wake behind the satellite of density much lower than found in the ambient plasma and further decreasing the current collection. These two effects mean that the collection of neutralizing ion currents to a

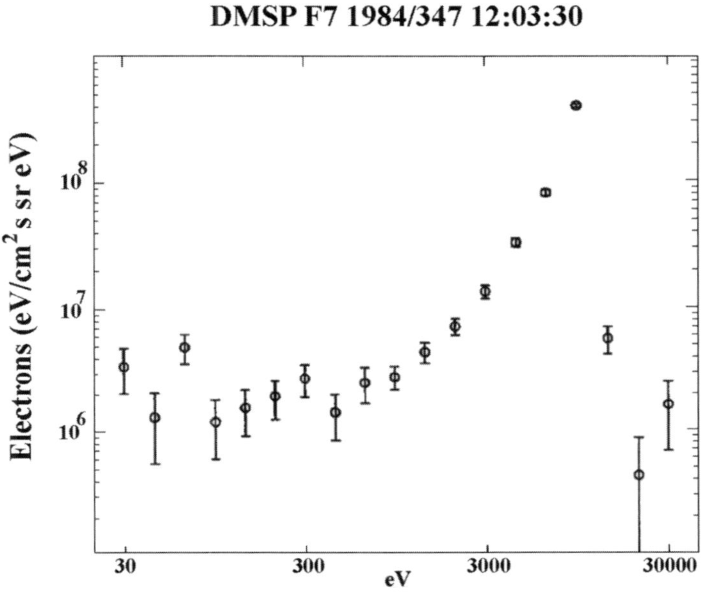

# DMSP F7 1984/347 12:03:30

FIG. 6.1   Electron spectra observed on the DMSP F13 satellite on 12 Dec. 1984 (figure adapted from Newell et al. [10]).

negatively charged satellite is not facilitated as much in LEO compared to GEO as the simple comparison of densities and temperatures might suggest.

The active agent behind charging in the auroral regions is energetic electrons, with energies in the range of 5–10 keV, precipitating along magnetic field lines with sufficient intensity to affect the spacecraft charging. Such electron fluxes, in an energy range where many materials have a low secondary electron yield, can certainly cause spacecraft charging [11]. Currents aligned to the magnetic field, flowing between the ionosphere and the magnetosphere, mainly drive auroral physics. Because of the convergence of the geomagnetic field lines, these currents show much higher current densities at low altitudes than higher up in the magnetosphere. Furthermore, the downward-flowing electrons carrying upward currents get accelerated by auroral electric fields, redistributing auroral electron flux to higher energies at lower altitudes. As a consequence, the directed fluxes of electrons in the kiloelectron-volt range on auroral field lines can be significantly higher in LEO than in the magnetosphere.

Satellites are not necessarily confined to LEO or GEO orbits. The intermediate region between these altitude ranges, sometimes called MEO, is frequently used. This is a challenging environment from the point of view of modeling of spacecraft charging. The reason is that the scale lengths and spacecraft sizes are comparable. Experimental results in this regime are rather limited. Many limitations on in situ space plasma instrumentation are caused more by perturbations of the spacecraft electrostatic environment than the instrument technology.

Thus, in auroral regions a spacecraft moving along its orbit could suddenly experience a transition from a situation with a spacecraft charging voltage no more than a volt or two negative, to the case where it is intercepting a high-energy stream of electrons. If this occurs in darkness, where photoemission is suppressed, and the electrons impinge on wake surfaces where ions are excluded by the spacecraft motion, then the potential for charging is high.

## 6.3   PLASMA EFFECTS ON SPACECRAFT IN THE AURORAL REGION

The space plasma interacts with space systems, leading to several effects including the following: surface charging, plasma anisotropies including wakes, ion impacts, current leakage between voltage generators, and interactions with plasma thruster plume. Surface charging occurs because electric charges (electrons and ions) of the plasma are free to move and eventually get collected on material surfaces when the charges hit them. They cannot leave a surface as easily as in the atmospheric environment because the vacuum is far less conductive regarding this process. Also, UV and soft-X photons coming from the sun have enough energy to expel electrons from materials hereby leaving an excess of positive charge.

The accumulation of charge (positive or negative) leads to the creation of an electric field that eventually will prevent further charge accumulation by repelling charges of a given polarity and attracting charges of opposite polarity. Indeed, the spacecraft is like a complicated electrical circuit with both active and passive elements that is coupled with electrical current in space via its whole surface.

The arcing induced by high voltage charging, which has been known for a long time to be a problem in a high-altitude orbit in a hot and low-density plasma environment (especially geostationary), also is a concern in the low-altitude high inclination orbit over the aurora.

## 6.4   AURORAL CHARGING IN MIDDLE EARTH ORBIT

### 6.4.1   FREJA STUDY

Very few studies have been devoted to spacecraft charging in PEO. One exception is the series of DMSP spacecraft at an altitude of around 840 km, which have been investigated in a suite of observational [e.g., 3, 5, 7] and numerical [e.g., 12, 13] studies. We will return to the DMSP results in Sec. V. The Swedish spacecraft Freja at roughly twice that altitude had comprehensive instrumentation that has further added to the knowledge on spacecraft charging in the PEO at intermediate altitudes. The large Freja database, special orbit, and comprehensive instrumentation made possible a series of studies of charging events, including detailed statistics and numerical simulations. Summaries of these studies have been produced by Koskinen et al. [14] and by Eriksson and Wahlund [15], to which we refer for further details.

Freja was launched on 6 October 1992 and was operated for four years, two years longer than the nominal lifetime. The first two years covered a declining phase of the solar cycle from medium to minimum solar activity. Freja was spin-stabilized and sun-pointing with a 6-s spin period. The orbit had 63-deg inclination, an apogee in the northern hemisphere of 1756 km, and a perigee in the southern hemisphere of 601 km. Freja therefore passed along the auroral region almost tangentially and can be classified as a PEO spacecraft despite its low inclination. In fact, the 63-deg inclination meant that Freja skimmed the auroral oval rather than just crossing it quickly, and thus spent longer time there than satellites at higher inclination, further adding to the suitability of Freja for auroral charging studies.

In the design of Freja, special care was taken to cover conductive surfaces with material of high secondary electron emission, mainly indium tin oxide (ITO), yet high levels of charging appeared. Measurement disturbances were observed in, for example, the Langmuir probe and plasma wave data, as well as in the energetic ion data. The fact that no other operational problems were observed during the charging events suggests that no electrostatic discharges occurred, presumably because of the lack of insulators adjacent to the high-conductivity spacecraft surface.

During charging events, the interpretation of the plasma measurements is often difficult. However, with careful analysis, it is possible to derive characteristics of the plasma environment leading to electrostatic charging.

Transverse ion heating produces ion distributions similar to those during spacecraft charging. However, the detailed ion energy pitch-angle distributions during these events are considerably different and can easily be distinguished. Also, transverse ion heating in the Freja data set is most often associated with field-aligned suprathermal electron bursts rather than isotropic inverted-V precipitation. Figure 6.2 shows an example of Freja data including effects from charging as well as transverse ion heating. The spacecraft charging is most easily seen in the three-dimensional ion composition spectrometer (TICS) oxygen ion data during the time interval from about 09.17.00 to 09.18.10. The electron data from the two-dimensional electron spectrometer (TESP) instrument show an increase in the energy range from a couple of kiloelectron volts to more than 10 keV in the same time interval, demonstrating a good correlation between the presence of auroral electrons and spacecraft charging. Poleward of that structure ion, data show effects of charging and transverse heating.

Surface charging events on the Freja spacecraft were defined using the following criteria:

• The Langmuir probe currents of the spherical as well as the cylindrical probe should be below $2 \times 10^{-8}$ A. This indicates that the plasma density is extremely low or the probes are negatively charged to a few volts. This resulted in about 500 possible events.

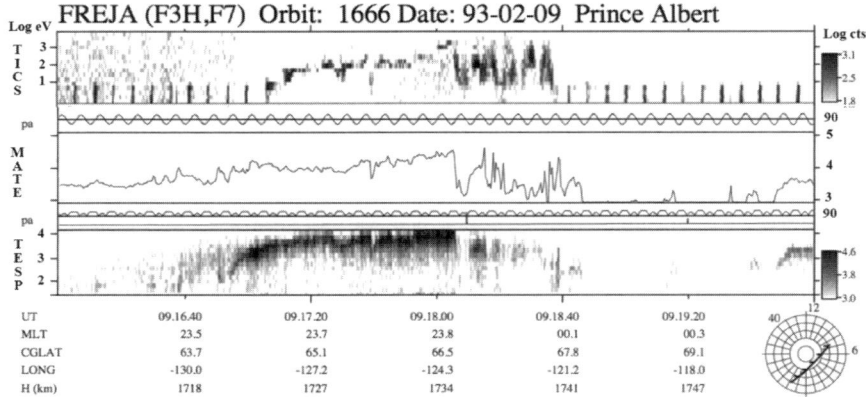

**FIG. 6.2  Example of the effect of spacecraft charging in the auroral zone as observed on Freja. The lowest panel shows electron fluxes in the energy range 0.01 to 20 keV, the middle panel integrated electron flux, and the uppermost oxygen ions 1 eV to 4.5 keV. The pitch angles (pa) covered by the electron and ion detectors are also shown.**

•   The narrowband Langmuir emission indicated a larger density than the Langmuir probe currents would indicate. During charging events, the narrowband emission is often unaffected, while the Langmuir probe currents drop to very low values, thus suggesting that the density is much larger than the Langmuir probe current would suggest, indicating negative charging.

•   The ion data showed a clear increase in energy of at least 5 eV. If the ion distributions have characteristics similar to transverse ion heating events (ion conics), they were only kept as suspect charging events if there were a clear mismatch between the densities inferred from the Langmuir probes and the narrowband Langmuir emissions.

All events with a charging level above about 5 V negative were kept in a database. The preceding procedure resulted in a total of 291 charging events.

   All charging events show some degree of intensification in the plasma wave emissions, from below heavy ion gyrofrequencies (typically 25 Hz) up to the plasma frequency range (hundreds of kilohertz). The narrowband Langmuir waves, and/or the upper cutoff of electrostatic whistler-type waves, gave estimates of the electron density. The electron density inferred in this way most of the time agreed well with the probe current measured by the Langmuir probes. However, during charging events the Langmuir probe current drops to very low values and is no longer a good estimate of the plasma density. In such cases the narrowband Langmuir emission is still an accurate tool for determining the electron density, and its discrepancy with the probe current becomes a good indicator of negative charging (see the following).

The plasma density can also be estimated from the low-frequency cutoff of whistler waves. This cutoff is usually around a few kilohertz. During a charging event, the electric field measurements in this frequency range are strongly disturbed, but the weak magnetic component of the whistler emission is often sufficient to determine this cutoff.

It also became clear why the Langmuir probe current could drop without any signature of density depletion in the wave data. The probe is positively biased with respect to the satellite, but when the spacecraft goes sufficiently negative, the probe will also be at negative potential with respect to the plasma. In this situation, the probe current drops, irrespective of the density but consistent with the observed charging.

A detailed description of all experiments onboard Freja can be found in a special Freja instrumentation issue of *Space Science Reviews* [16].

## 6.4.2  CHARACTERISTICS OF FREJA CHARGING EVENTS

A study of the 291 Freja events showed four different types of charging: 1) charging by energetic electrons during eclipse, 2) charging by energetic electrons during sunlight and terminator conditions, 3) low-level charging variations due to sunrise/sunset, and 4) low-level charging probably due to an increased bulk electron temperature.

The first category of events is by far the most common type of high-level charging. The two last usually result in charging levels below the 5-V threshold that was chosen for the statistical study, and therefore only very few such events were included.

The main part of the charging events reached between 10 and 100 V (negative). Only a handful of the events reached charging levels around −2000 V. The peak energy of the precipitating energetic electrons and their high-energy tail seemed to control the charging level during the high-level events.

Spacecraft charging to low levels (up to 10 eV negative) sometimes occurred without simultaneous intense energetic electron fluxes. One explanation for this behavior can be that Freja entered regions with high electron temperature and rather low density. However, it was very seldom that low-level charging above a few volts was detected in the Freja data set without simultaneous energetic electron precipitation, and electrons above a few kiloelectron volts is certainly the most dominant source of Freja charging events. Low-level charging was detected when enhanced electron fluxes below an energy of 10 keV occurred. This is an important result because Freja is to the largest extent covered with ITO that has a second crossover energy (the energy at which the secondary yield equals one) of 2.5–3 keV, and the second most common surface material is a thermal blanket with crossover energy of just under 4 keV. Thus, the high-energy tail (10–80 keV) is not necessary for charging in these observed cases. These charging events can be explained by a large electron temperature (2–3 eV) of the thermal plasma. However, for charging to more than 100 eV negative, peak energies

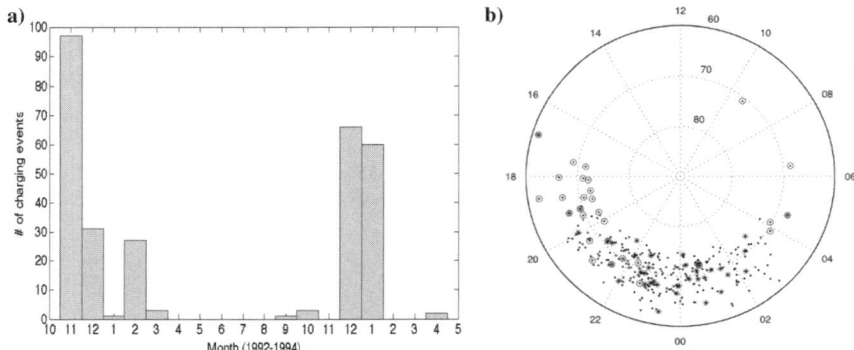

**FIG. 6.3** Distribution of Freja charging events with a) time of year and b) in space. All events are from the northern hemisphere for reasons of ground-station coverage. The plot in Fig. 6.3b uses corrected geomagnetic coordinates. Dots and stars mark charging events to levels less or more negative than −100 V, respectively. Rings mark events in sunlight.

usually were around 10 keV, and there was a significant high-energy tail in the tens of kiloelectron-volts range, similar to the spectrum shown in Fig. 6.1.

No charging events occurred during the summer months, but a maximum occurred in the winter months (see Fig. 6.3). This behavior is most probably caused by the combined effect of sunlight (photoelectron emissions) and the expected larger thermal plasma densities due to upward diffusion from the more pronounced ionospheric F2 layer during the summer months. Most charging occurred in the MLT interval 1800–0300 hours and in a latitude band consistent with the statistical auroral oval as shown in Fig. 6.3.

### 6.4.3  MODELING OF THE FREJA EVENTS

The plasma density in the Freja charging events varied between 20 and 2000 cm$^{-3}$. The electron temperature was assumed to stay between 0.2 and 2 eV. In these circumstances, the electron Debye length varies from less than 10 cm to over 2 m. Given a magnetic field of 25–30 μT, the electron gyroradius is on the order of 10 cm, and the cold ion gyroradius is on the order of 5 m for protons and 20 m for oxygen ions. Freja is 2.2 m in diameter and around 50 cm in height. The spacecraft dimensions are thus on the same order as several characteristic dimensions of the plasma, necessitating the use of numerical simulations for understanding the complex processes involved in charging.

Some engineering models of spacecraft charging, including the NASA Spacecraft Charging Analysis Tool (NASCAP) and the U.S. Air Force (USAF) code POLAR, could not easily reproduce the high-level charging observed with the Freja plasma measurements [15, 17]. The POLAR (Potential of Large Spacecraft in the Auroral Region) code package, described by Cooke [13], had previously been successfully applied to auroral charging events on the DMSP satellites.

Several Freja models were used in the simulations. The most detailed model, capable of including localized exposed insulators like telemetry antenna covers, had a resolution of about 10 cm. As POLAR uses a fixed grid, of equal resolution for spacecraft and space, simpler models were used for trial simulations.

Five of the Freja charging events were modeled. Observed electron spectra were used in POLAR for the description of auroral electrons. It was necessary to correct the observed electron spectra for the charging level determined from the ion data. Charging to moderate levels ($-50$ V) could be successfully modeled. Any significant charging could not be obtained in sunlight if photoemission was included in the simulation. However, to reproduce the kilovolt potentials seen in some of the events, the environmental parameters and/or materials constants had to be varied outside the range supported by the data.

The POLAR results for Freja were compared with identical environment parameters and electron spectra, taken from a DMSP charging event. For an environment where POLAR gave $-195$ V for DMSP, Freja only charged to $-40$ V. This was expected, given the smaller size and different surface materials of Freja.

There are many possible reasons for the inability to model the highest charging levels with POLAR. Material properties is one area of uncertainty, particularly if some materials change in space, though the appearance of charging events within a month after launch would require such change to be quick. Quenching of photoelectrons or secondary electrons due to barrier formation can possibly occur. For the sunlit cases, the large solar zenith angle serves to decrease UV intensity and hence photoemission. In addition, decreased photoemission by reflectance effects is a possibility [18]. As shown by the explanation of its acronym, POLAR was developed primarily for large spacecraft, that is, structures significantly larger than the Debye length. Freja is smaller than DMSP, and the plasma is less dense. In the simulations, the sheath tended to expand to the edge of the simulation box, which might indicate a problem here. Finally, the field-aligned electron fluxes can be underestimated because of the limited angular coverage of the Freja electron detectors. Despite the lack of detailed reproduction of observed charging levels through simulations, the Freja results are in qualitative agreement with standard surface charging theory [e.g., 12, 19–21].

## 6.4.4  DEPENDENCE ON SUNLIGHT CONDITIONS

Of the 291 Freja charging events, 32 occurred during sunlight conditions, 236 during eclipse, and 23 during terminator conditions. The large fraction of the events seen in sunlight might seem surprising, as spacecraft photoemission here should counteract the negative charging of the spacecraft. However, all of the sunlight events were found during northern winter (no southern hemisphere data were included in the Freja study) and above a dark (nightside) ionosphere, both implying lower local plasma density. The photoemission current was thus counteracted by the lower ion collection current in these events. Finally, Freja was operating at solar minimum, when the solar UV intensity responsible for photo-emission has low intensity, which has been shown previously to be the period when charging most likely occurs [5].

### 6.4.5  DEPENDENCE ON GEOMAGNETIC LOCATION AND GEOMAGNETIC ACTIVITY

Freja charging events occurred during nighttime hours with an event peak around 2200–2300 magnetic local time (MLT; Fig. 6.4). Almost no charging events existed between 0600–1800 MLT. Thus most charging events occur in the absence of the photoelectron emission from the spacecraft. All MLT values were covered rather evenly by the Freja orbits.

The Freja events were binned with the three-hour averages of the Kp index, an indicator for geomagnetic activity. This showed that there was a weak increasing tendency with increasing Kp indices and that the probability of charging was large for Kp greater than 2+.

The Kp index dependence for the Freja charging events is plotted in Fig. 6.5. It shows that 1) low Kp events tend to occur in a narrow local time sector near local midnight; 2) high Kp events can happen during a broader local time sector on the nightside, although the Freja spacecraft need not be in shadow; 3) low Kp events occur only during eclipse; and 4) no obvious dependence with altitude exists, even though a weak trend favors lower altitudes for larger Kp indices.

### 6.4.6  DEPENDENCE ON PLASMA CHARACTERISTICS

The electron flux and peak energy are major factors determining the charging level for a given plasma density. During the charging event, the maximum flux of energetic electrons at the peak moves above the crossover energies of the dominating surface materials, ITO (2.5–3 keV) and the thermal blanket (just below 4 keV). This suggests that the flux of the emitted secondary electrons cannot balance the incident energetic electron flux longer and in this way causes the

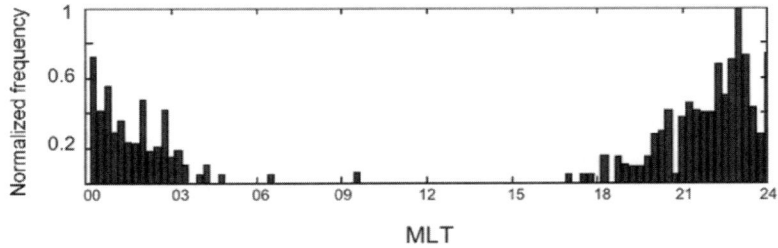

**FIG. 6.4  MLT distribution of charging events seen by the Freja spacecraft.**

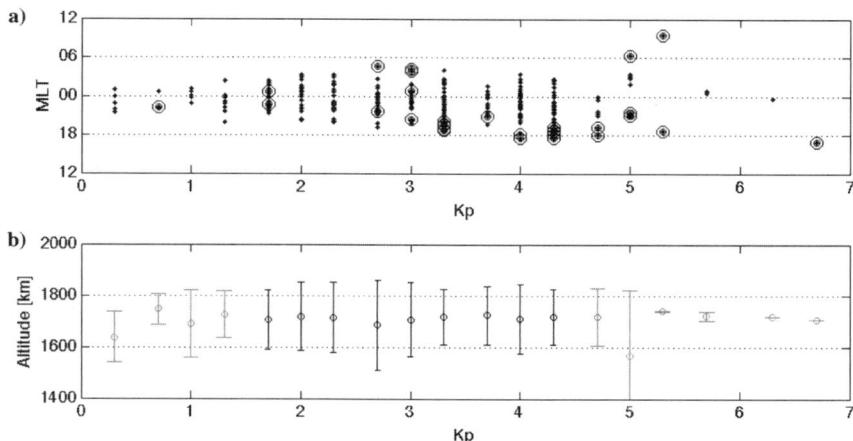

**FIG. 6.5    Kp dependence of Freja charging events: a) distribution in MLT as function of Kp where circles identify sunlit or terminator events; b) event altitude variation with Kp (error bars at plus minus two standard deviations) where gray bars mark Kp values for which less than 10 events were found.**

observed negative charging. Also, the flux levels in the high-energy tail increase by about an order of magnitude, which further contribute to the accumulation of negative charge.

There was an almost one-to-one correlation between charging events and energetic electron flux (as well as energy) increases in the Freja database. Even the low-level charging events of a few volts negative were associated with the presence of energetic electrons. Typical flux levels at the inverted-V energy peak during single charging events were in the range $10^6$–$10^8$ $(cm^2$-s-str-keV$)^{-1}$.

The Freja data show an upper threshold of 2 x $10^9$ m$^{-3}$ for the cold plasma density. The thermal plasma density seldom decreased below $10^8$ m$^{-3}$. This can be compared with the results obtained by Frooninckx and Sojka [5], who detected a similar electron density threshold value of $10^{10}$ m$^{-3}$ for the DMSP satellites. The DMSP satellites traversed lower altitudes (840 km) and were therefore in regions with larger thermal plasma densities. However, the DMSP satellites are also much bigger than Freja, creating a deeper wake behind them, and have a larger fraction of their exposed surface area covered by insulating materials. These factors both contribute to DMSP experiencing charging in a wider density range than Freja, despite the higher ion fluxes expected at DMSP altitude.

Three cases of Freja charging with rather similar inverted-V electron spectra revealed that the ambient plasma density affects the charging levels. A change in density by a factor 5–8 seemed to be associated with a change in charging level by a similar factor 5–10.

The influence of low-energy suprathermal electron bursts on charging level is not obvious in the data, but a high-energy electron tail seemed to enhance the charging levels. High fluxes of megaelectron-volt electrons might be associated with the maximum charging part of some charging events. Charging occurs only when the peak energy of the inverted-V electron population reaches above about 5 keV, although no detectable charging exists for lower inverted-V peak energies.

To summarize, a high-energy electron flux level is indeed needed, but other physical parameters, such as the thermal plasma density and the ion return current from this plasma to the spacecraft as well as the low-energy electron flux, which produces many secondary electrons, determine the charging level as well. The Freja results are in qualitative agreement with standard surface charging theory [e.g., 12, 19, 20].

### 6.4.7    DEPENDENCE ON SIZE OF THE SPACECRAFT, ALTITUDE, AND MAGNETIC FIELD STRENGTH

In the Freja study the normalized altitudes of charging events were rather evenly distributed. An increasing trend could be detected with increasing altitude, but this trend was within the possible errors. In low Earth orbit, the geomagnetic field **B** is strong enough so that secondary electrons and photoelectrons emitted from the spacecraft surface have an average gyroradius smaller than typical dimensions of a spacecraft.

The gyroradius at PEO/LEO altitudes for 2-eV electrons is 0.1–0.4 m. Some of the secondary electrons produced on the surface will return to the spacecraft because of the gyromotion and the orientation with respect to the Earth's magnetic field direction. For a larger spacecraft a larger fraction of the secondary electrons will return to the spacecraft surface. A larger spacecraft will therefore most likely reach somewhat higher charging levels. The same argument is true with regard to sunlight conditions, when large amounts of photoelectrons are emitted from the surface. A smaller spacecraft will not see the photoelectrons return as easily as a larger spacecraft, and the larger spacecraft is therefore more easily charged.

For Freja the gyroradius and dimensions were comparable. This implies that escape of secondary electrons decreases on surfaces nearly parallel with **B**, which in turn can affect the current balance of the spacecraft and make high-voltage charging more likely [22]. Nevertheless, the magnetic field orientation could not be shown to be a major factor for the charging level of the Freja satellite. For a perfectly spherical spacecraft, such a result is expected irrespective of gyroradius. Although Freja is not a perfect sphere (it can more reasonably be described as a cylinder or an oblate ellipsoid of major and minor axes 2 and 0.5 m, respectively), the variation in secondary emission with B-field direction for this geometry can be concluded not to be sufficiently strong to be obvious in the charging statistics.

## 6.4.8  TIME DURATION OF CHARGING EVENTS

Most of the Freja charging events were shorter than one minute, but a few lasted for a couple to several minutes. Only very few continued for several minutes even though Freja had an orbit that passed almost tangentially along the auroral oval. For charging levels above 10 V, there seemed to be a preferred duration of about 20 s. The duration of charging events reflects the time it takes for Freja to traverse the auroral structures, that is, the high-energy electron precipitation region. Thus, when the source for charging disappears, the charging also disappears.

It was therefore concluded that very little differential charging between Freja surfaces occurred, which was also confirmed by the POLAR simulations.

## 6.5  AURORAL CHARGING IN LOW EARTH ORBIT

The largest database on spacecraft charging at low altitudes in the auroral zone was, and still is, collected by the DMSP satellites [7]. Although the charging in LEO in many respects is similar to what was found in MEO above (e.g., charging to large negative potentials in regions of high kiloelectron-volt electron fluxes), there are a few observational facts that distinguish the DMSP results at lower altitude from the Freja observations already discussed. Some of these differences have already been discussed.

The DMSP charging events were colocated with plasma depletions, for example, [3], although the Freja results indicated that thermal plasma density is often largely unaffected. The electron current to positively biased Langmuir probes was indeed found to almost disappear in a charging event, but this proved not to be a reliable indicator of the plasma density in these circumstances. On Freja, the density could be determined in two independent ways from the Langmuir probes and from the plasma frequency. The plasma density determined from the plasma frequency often showed little sign of depletion during a charging event, and so the decreased current to the Langmuir probes was simply an effect of the negative spacecraft potential expelling plasma electrons from the vicinity of the satellite.

Consequently, the correlation between charging events and unusually low densities is much clearer on DMSP than on Freja. This can be understood from the fact that densities at Freja orbit (1700 km) are generally much lower than what DMSP encounters at roughly half the altitude. The plasma density values encountered in DMSP charging events are low for that altitude, but not at Freja orbit, and so further depletion of the plasma is not a necessary condition for charging in MEO. In LEO, densities are normally so high as to inhibit charging even in the presence of tens of kiloelectron-volt electron fluxes that would otherwise be sufficient for causing spacecraft charging. Frooninckx and Sojka [5] found a better correlation with the DMSP charging level if they divided the high-energy electron flux by the ambient plasma density.

Electrons with as low energy as 2–3 keV contributed to charging onboard DMSP [5], whereas energies above 5 keV were necessary to charge Freja. Also, the upper-threshold plasma density making surface charging possible onboard DMSP was about $10^{10}$ m$^{-3}$, while the Freja threshold was a factor of five lower ($2 \times 10^9$ m$^{-3}$). This reflects the different surface material properties of the two spacecraft with respect to charging and the effect of the larger size of DMSP (compare Sec. 6.4.6).

Another difference between Freja and DMSP is the 30 events occurring in sunlight, of which seven reached potentials more negative than $-100$ V. From DMSP, no charging event was found in sunlight. However, although Freja is sunlit at these events, the ionosphere below is in darkness. This is consistent with auroral acceleration occurring mainly over the dark ionosphere [23, 24].

The DMSP results can be summarized as follows:

1.  High-latitude charging events were detected in association with intense auroral energetic electron precipitation colocated with ionospheric plasma depletions (e.g. [3]).

2.  Precipitating electrons with energies as low as 2–3 keV can contribute to surface charging, although higher-energy electrons make a greater contribution [5].

3.  Incident energetic electron fluxes of at least $10^{12}$ electrons/m$^2$-s-str existed during the charging events [3].

4.  The charging level was better correlated with the ratio of the integral electron flux over the cold thermal plasma density rather than just the integral electron flux alone [4].

5.  Secondary electron production at the surface by the incident electrons with energies below 1 keV inhibited charging, most probably because of the large secondary yields at these low incident electron energies [5].

6.  The charging events always occurred during eclipse, that is, in the absence of photoelectron emission from the spacecraft surface [3].

7.  The highest thermal plasma density observed during charging events was $10^{10}$ m$^{-3}$ [5].

8.  The distribution of the charging events was modulated by the solar activity, that is, the charging was most frequent and severe during solar minimum. Because the incident electron flux did not vary significantly over a solar cycle, although the cold thermal plasma density varied with several orders of magnitude, it was suggested that the increased ionization, and resulting increased thermal plasma density, during solar maximum inhibited charging [5].

9.  Theoretical comparisons with the POLAR code suggested, except for the material secondary electron yield properties, that the spacecraft orientation,

fraction of conductive areas in the velocity direction, the cold thermal plasma density, and the ion composition modified the charging level to the greatest extent [6].

## 6.6   DISCUSSION AND CONCLUSIONS FOR FUTURE MISSIONS

Electrical charging of the surface and internal structure of spacecraft have been found to cause operational problems to spacecraft as well as to their onboard instrumentation. When the resulting electric field exceeds the breakdown field intensity for a material, an electrostatic discharge occurs. Even charging to only a few tens of volts negative can cause serious operational effects in certain instruments.

When crossing the auroral oval, high-level surface charging is likely to continue to occur on spacecraft in PEO/LEO, unless serious engineering efforts are made to produce surface materials with large secondary yields for incident electron energies in excess of tens of kiloelectron volts. Active methods to reduce charging levels exist, like expelling plasma from the spacecraft [25], but such methods can increase the contamination and differential charging problems.

Although the Freja spacecraft was mainly coated with materials of good secondary emission properties, the spacecraft electrostatic potential went down below 1 kV negative on some occasions. Surface charging should therefore be of concern for future spacecraft designs, particularly if the spacecraft surface will include exposed insulators, liable to differential charging, or if there is a need to keep a stable and low spacecraft potential for example, for scientific measurements. The analysis should be complemented with precise electron and ion-energy/pitch-angle distributions, sunlight/eclipse characteristics, geomagnetic location, and auroral activity conditions, as well as cold plasma information. This is important not only around Earth: predictions for spacecraft charging in the auroral regions at Jupiter and Saturn indicate that charging is also an issue for future space missions to the outer planets [26].

The orbit of Freja in the intermediate-altitude region between the well-studied LEO and GEO altitude ranges is underrepresented in spacecraft charging studies.

To predict the charging level of a LEO/PEO spacecraft, one needs to know precisely the incident electron and ion spectral distribution as well as the surface material properties (especially the secondary yield vs incident energy) because the current balance is largely determined by the ratio between the flux of the incident electrons and the flux of reemitted secondary electrons (e.g., see [1, 4] and also [26] regarding similar ATS and SCATHA results).

The Freja study noted that the electron density derived from a Langmuir probe in electron collection mode was not reliable during charging events because of electron repulsion from the negatively charged spacecraft. In general, measurement reliability during charging conditions must be taken into account when analyzing charging data.

The Freja satellite was designed to be highly conductive and as electromagnetically clean as possible and therefore should be better prepared for charging effects compared to the DMSP satellites. Yet, the spacecraft achieved somewhat larger maximum charging levels and also charged during sunlight conditions, which the DMSP satellites never did [3]. This is mainly explained by the lower plasma densities at Freja altitudes, but the Freja mission occurred close to solar minimum, which has been proven to be the period when charging is most likely to occur [5].

## REFERENCES

[1]   Katz, I., and Parks, D. E., "Space Shuttle Orbiter Charging," *Journal of Spacecraft and Rockets*, Vol. 20, No. 1, 1983, pp. 22–25.

[2]   Hall, W. N., Leung, P., Katz, I., Jongeward, G. A., Lilley, J. R., Jr., Nanevicz, J. E., Tayler, J. S., and Stevens, N. J., "Polar-Auroral Charging of the Space Shuttle and EVA Astronaut," *The Aerospace Environment at High Altitudes and Its Implications for Spacecraft Charging and Communications*, edited by E. R. Schmerling, AGARD-CP-406, North Atlantic Treaty Organization, France, 1987, pp. 34.1–34.9.

[3]   Gussenhoven, M. S., Hardy, D. A., Rich, F., Burke, W. J., and Yeh, H.-C., "High-Level Spacecraft Charging in the Low-Altitude Polar Auroral Environment," *Journal of Geophysical Research*, Vol. 90, No. A11, 1985, pp. 11009–11023.

[4]   Yeh, H.-C., and Gussenhoven, M. S., "The Statistical Electron Environment for Defense Meteorological Satellite Program Eclipse Charging," *Journal of Geophysical Research*, Vol. 92, No. A7, 1987, pp. 7705–7715.

[5]   Frooninckx, T. B., and Sojka, J. J., "Solar Cycle Dependence of Spacecraft Charging in Low Earth Orbit," *Journal of Geophysical Research*, Vol. 97, No. A3, 1992, pp. 2985–2996.

[6]   Stevens, N. J., and Jones, M. R., "Comparison of Auroral Charging Predictions to DMSP Data," AIAA Paper 95-0370, Jan.1995.

[7]   Anderson, P. C., "Spacecraft Charging Hazards in Low Earth Orbit," *Proceedings of the 9th Spacecraft Charging Technology Conference* [CD-ROM], edited by T. Goka, JAXA-SP-05-001E, Japan Aerospace Exploration Agency, Tsukuba Space Center, Ibaraki, Japan, 2005, pp. 813–825.

[8]   Kawakita, S., Kusawake, H., Takahashi, M., Maejima, H., Kurosaki, T., Kojima, Y., Goto, D., Kimoto, Y., Ishizawa, J., Nakamura, M., Kim, J.-H., Hosoda, S., Cho, M., Toyoda, K., and Nozaki, Y., "Investigation of an Operational Anomaly of the ADEOS-II Satellite," *Proceedings of the 9th Spacecraft Charging Technology Conference* [CD-ROM], edited by T. Goka, JAXA-SP-05-001E, Japan Aerospace Exploration Agency, Tsukuba Space Center, Ibaraki, Japan, 2005.

[9]   Nakamura, M., "Space Plasma Environment at the ADEOS-II Anomaly," *Proceedings of the 9th Spacecraft Charging Technology Conference* [CD-ROM], edited by T. Goka, JAXA-SP-05-001E, Japan Aerospace Exploration Agency, Tsukuba Space Center, Ibaraki, Japan, 2005.

[10]  Newell, P. T., Lyons, K. M., and Meng, C.-I., "A Large Survey of Electron Acceleration Events," *Journal of Geophysical Research*, Vol. 101, No. A2, 1996, pp. 2599–2614.

[11] Hastings, D. E., "A Review of Plasma Interactions with Spacecraft in Low Earth Orbit," *Journal of Geophysical Research*, Vol. 100, No. A8, 1995, pp. 14,457–14,483.

[12] Cooke, D. L., Gussenhoven, M. S., Hardy, D. A., Tautz, M., Katz, I., Jongeward, G., and Lilley, J. R., "Polar Code Simulation of DMSP Satellite Auroral Charging," *Proceedings of the Spacecraft Charging Technology Conference*, edited by R. C. Olsen, PL-TR-93-2027(I), Naval Postgraduate School, Monterey, CA, 1993, pp. 194–203.

[13] Cooke, D. L., "Simulation of an Auroral Charging Anomaly on the DMSP Satellite," *Proceedings of the 6th Spacecraft Charging Technology Conference*, AFRL-VS-TR-20001578, Air Force Research Lab., Hanscom AFB, MA, 2000, pp. 33–37.

[14] Koskinen, H., Eliasson, L., Holback, B., Andersson, L., Eriksson, A., Mälkki, A., Norberg, O., Pulkkinen, T., Viljanen, A., Wahlund, J.-E., and Wu, J.-G., "Space Weather and Interactions with Spacecraft," Technology Research Programme, ESTEC, Technical Rept. ESTEC/Contract No. 11974/96/NL/JG(SC), Noordwijk, The Netherlands, 1999, http://www.ava.fmi.fi/spee/.

[15] Eriksson, A. I., and Wahlund, J.-E., "Charging of the Freja Satellite in the Auroral Zone," *IEEE Transactions on Plasma Science*, Vol. 34, No. 5, 2005, pp. 2038–2045.

[16] Lundin, R., Haerendel, G., and Grahn, S. (eds.), "The Freja Mission," *Space Science Reviews*, Vol. 70, Nos. 3–4, 1994.

[17] Svensson, U., "Simulation of Spacecraft Charging in the Aurora: a Case Study," Space Systems Environment Analysis Sec., Mathematics and Software Div., ESA/ESTEC, Internal Working Paper 1943, Noordwijk, The Netherlands, July 1997.

[18] Lai, S. T., and Tautz, M. F., "Aspects of Spacecraft Charging in Sunlight," *IEEE Transactions on Plasma Science*, Vol. 34, No. 5, 2006, pp. 2053–2061.

[19] Garrett, H. B., "The Charging of Spacecraft Surfaces," *Reviews of Geophysics and Space Physics*, Vol. 19, No. 4, 1981, pp. 577–616.

[20] Hastings, D., and Garrett, H. (eds.), *Spacecraft Environment Interactions*, Atmospheric and Space Science Series, Cambridge Univ. Press, New York, 1996, p. 292.

[21] Garrett, H. B., and Whittlesey, A. C., "Spacecraft Charging: an Update," *IEEE Transactions of Plasma Science*, Vol. 28, No. 6, 2000, pp. 2017–2028.

[22] Laframboise, J. G., "Calculation of Escape Currents of Electrons Emitted from Negatively Charged Spacecraft Surfaces in a Magnetic Field," *Journal of Geophysical Research*, Vol. 93, No. A3, 1988, pp. 1933–1943.

[23] Newell, P. T., Meng, C.-I., and Lyons, K. M., "Suppression of Discrete Aurorae by Sunlight," *Nature*, Vol. 381, 27 June 1996, pp. 766, 767.

[24] Hamrin, M., André, M., Norqvist, P., and Rönnmark, K., "The Importance of a Dark Ionosphere for Ion Heating and Auroral Arc Formation," *Geophysical Research Letters*, Vol. 27, No. 11, 2000, pp. 1635–1638.

[25] Usui, H., Imasato, K., and Kuninaka, H., "Three-Dimensional Particle-in-Cell Simulations on Active Mitigation of Spacecraft Charging in the Earth's Polar Region," *Proceedings of the 26th International Symposium on Rarefied Gas Dynamics*, edited by T. Abe, AIP Conference Proceedings, Vol. 1084, Japan Aerospace Exploration Agency, Tsukuba, Japan, 2008, pp. 877–882.

[26] Garrett, H. B., Evans, R. W., Whittlesey, A. C., Katz, I., and Jun, I., "Modeling of the Jovian Auroral Environment and Its Effects on Spacecraft Charging," *IEEE Transactions on Plasma Science*, Vol. 36, No. 5, 2008, pp. 2440–2449.

[27] Olsen, R. C., "A Threshold Effect for Spacecraft Charging," *Journal of Geophysical Research*, Vol. 88, No. A1, 1983, pp. 493–499.

# CHAPTER 7

# Internal Charging

D. J. Rodgers* and J. Sørensen*

*ESA/ESTEC, Noordwijk, The Netherlands*

## 7.1   INTRODUCTION

The outer radiation belt of the Earth contains a population of energetic electrons (several hundred kiloelectron volts and above) that have enough energy to penetrate the outer surface of a spacecraft and that therefore interact with internal materials. This outer belt extends from MEO to GEO altitudes and therefore affects, for example, MEO navigation spacecraft and geostationary telecommunications satellites. High fluxes of electrons with similar energies are also found in the trapped radiation belts of Jupiter and Saturn.

High-energy electrons are scattered as they pass though spacecraft materials, depositing energy (i.e., radiation dose) and slowing down as a result. This process injects charging current where the electrons finally come to rest. Higher-energy particles have a greater probability of penetrating more deeply into the spacecraft than lower-energy particles. Because lower-energy particles are more plentiful, the charging currents are greatest in materials on or just under the spacecraft surface.

Current can be deposited both in internal metals and dielectrics and both deep within the spacecraft and within surface layers. We call this process "internal charging," but the names "bulk charging," "deep-dielectric charging," "internal dielectric charging," and "thin-film charging" are widely used to describe essentially the same phenomenon.

Internal charging currents to conductors linked to spacecraft ground make a negligibly small contribution to the overall current to spacecraft ground in comparison to other environmental factors, such as photoelectron emission and plasma current. However, floating (i.e., ungrounded) metals underneath the spacecraft surface are not exposed to other environmental interactions, and the internal charging current dominates their charging state. Similarly, currents in highly insulating dielectrics are dominated by internal charging, except for the

---

*Analyst, Space Environment and Effects, Keplerlaan 1, 2201.

outermost layers of surface dielectrics. Internal charging currents in such cases are trapped and cause a rise in the potential difference between floating metallic and dielectric structures and the rest of the spacecraft.

In contrast to surface charging, in which hazardously high potentials can be reached in seconds or minutes, the lower currents from internal charging generally take days or longer to reach hazardous levels. However, the consequence of charging can be the same in both cases—electrostatic discharge, with the risk of damaging effects on electronic equipment.

This chapter is intended to give an overview of the internal charging process, its effects, and its assessment.

## 7.2  MECHANISM

After energetic electrons penetrate the spacecraft surface, they undergo repeated scattering in which they lose energy. This is a material-dependent process characterized by the stopping power $S$, where

$$S(E) = -\frac{dE}{dx}$$

where $E$ is energy and $x$ is the path length.

The energy loss process includes the emission of bremsstrahlung photons and the ejection and subsequent reabsorption of secondary electrons. Within materials these interactions cause no net change in electric charge. At the surfaces, where secondary electrons might escape, some electrons might be lost, but these are only a small fraction of the incident number (i.e., the net secondary yield is well below unity for primary electrons of this energy). Hence, the main current is due to the capture of the primary electrons.

The ratio of electron penetration distance to material density is characterized statistically by the range $R$, and, over the years, various formulas have been proposed to describe this as a function of energy, including the following by Weber [1],

$$R = 0.55 \times E \times \left[1 - \frac{0.9841}{1 + 3E}\right] \text{ g cm}^{-2}$$

This formula has shown excellent agreement with Monte Carlo simulations performed with the ITS [2] and Geant-3 [3] codes [4]. Sorensen [5] observed from simulations using the ITS that an incident monoenergetic electron beam for the energies of interest exhibits a nearly linear decrease in intensity over a distance $a$ where

$$a = 0.238 \times Eg \text{ cm}^{-2}$$

This means that electrons with the same energy and direction are not all stopped at the same depth but over a range of depths, as illustrated in Fig. 7.1. This effect is called *straggle*. In comparison with ITS simulations, this equation

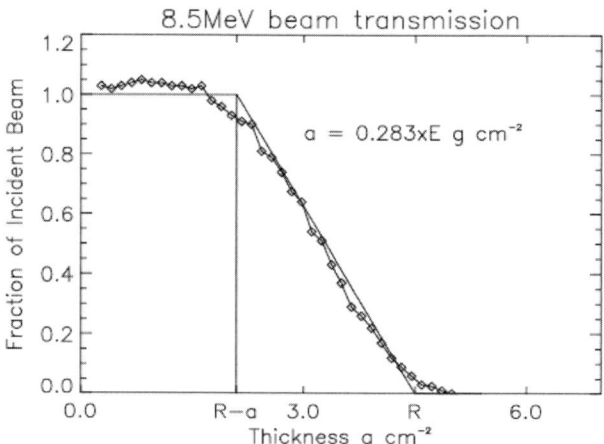

**FIG. 7.1   ITS simulation of electron-beam intensity in aluminum and modeled intensity combining the Weber [1] and Sorensen [5] formulas.**

has been shown to be a reasonable approximation (within 10%) for aluminum and Plexiglas® between 1 and 10 MeV [6].

An alternative analytical method for calculating current deposition as a function of depth was created by Frederickson and Bell [7], who modified the EDEPOS dose-depth algorithm [8] and fitted the new function to experimental measurements of charge deposition.

In an ungrounded metal, internal fields must be zero, and so internally deposited charge immediately moves to the surface, but in dielectrics this charge is, to a first approximation, trapped. The electric field due to electrical charge is described by Gauss' law.

$$\oint_S E.dA = \frac{Q}{\varepsilon}$$

where $E$ is electric field, $A$ is area, $Q$ is charge, and $\varepsilon$ is the permittivity. The electric field inside the dielectric depends on the dielectric geometry. The potential at a certain location relative to ground is the integral of the electric field between that location and the ground.

In a perfect insulator the charging continues until the electric field or surface potential is high enough to initiate an electrostatic discharge through dielectric breakdown or surface blowoff. Typically, a few tens of megavolts per meter or a few kilovolts surface potential are needed to trigger ESD.

In practice, even good insulators exhibit some conductivity, and this limits the charging to no more than the equilibrium situation in which as much current is conducted away as is deposited from the environment. The corresponding potential and electric field are described by Ohm's law.

$$V = IR \quad \text{or, equivalently,} \quad E = \frac{j}{\sigma}$$

where $V$ is voltage, $I$ is current, $R$ is resistance, $j$ is current density, and $\sigma$ is conductivity.

The time taken to reach equilibrium depends on the capacitance and resistance of the structure. In the simple case of a planar structure with uniform material properties, this resembles the charging of a planar capacitor, that is,

$$V = IR\left(1 - \exp\frac{-t}{\tau}\right) \quad \text{or equivalently} \quad E = \frac{j}{\sigma}\left(1 - \exp\frac{-t}{\tau}\right)$$

where $t$ is time and the time constant $\tau = \epsilon/\sigma$. This can be rewritten as $\tau = E_\infty \epsilon/j$, where $E_\infty$ is the equilibrium electric field. Here $\epsilon$ does not vary greatly between dielectrics (typically 1 to $5 \times 10^{-11}$ F m$^{-1}$), and $j$ is typically below 0.1 pA/cm$^2$ under the spacecraft's outer skin in the Earth's radiation belts. In addition, electric fields above $10^7$ V/m are hazardous. Therefore, it can be seen that only materials with time constants above a certain value ($>10^5$ s$\approx$1 day) experience hazardous charging levels in the Earth's radiation belts.

## 7.3 EFFECTS OF INTERNAL CHARGING

The main hazardous effect of internal charging is electrostatic discharge (ESD). Because internal charging can occur below the spacecraft skin, ESD can occur inside a spacecraft's Faraday cage, for example, in electrical connectors or printed circuit boards. This can quite easily lead to the direct injection of large transient currents into electronic circuits or the indirect production of transient currents through electromagnetic coupling. An example of ESD transients from a realistic laboratory experiment [9] is shown in Fig. 7.2. Such transients can cause temporary upsets or permanent damage in electronic components.

Additionally, breakdown leads to the production of tracks within the insulator as shown in Fig. 7.3, which can cause a permanent change in the material properties and cause the material to be degraded as an insulator. These tracks, called Lichtenburg figures, are visible in transparent materials and have a tree-like appearance.

Published breakdown electric fields for dielectrics are typically above $10^7$ V/m; however, in laboratory experiments [6] pulsing has sometimes been seen at lower macroscopic field strengths, perhaps because of enhanced fields at edges. For design purposes, a value of $10^6$ V/m is a reasonable "danger-level" that should not be exceeded.

Internal charging electric fields can also directly affect equipment that uses sensitive control of electric fields to perform actions or to make measurements. Examples are microelectromechanical systems (MEMS) [10, 11] and accelerometers using test masses, for example, as used in the LISA spacecraft [12]. In

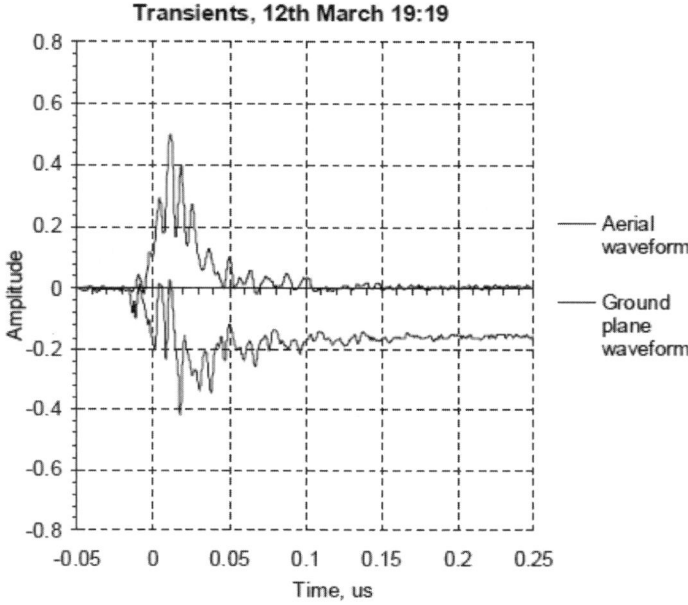

**FIG. 7.2  Discharge measurement of the directly injected current on an equipment ground plane (top trace) and the induced signal seen on a nearby antenna (bottom trace) [9]. Amplitude units are milliamperes.**

**FIG. 7.3  Lichtenberg figure in a transparent acrylic block after electron irradiation.**

the latter case, charging is most likely to be caused by highly penetrating cosmic rays because test masses are usually very strongly shielded against electrons.

## 7.4  SPACECRAFT ANOMALIES

It is often the case that the cause of serious spacecraft anomalies remains uncertain. In addition, owners of a spacecraft are sometimes reluctant to publish details on anomalies. However, evidence for internal charging generated discharges has been obtained both in the laboratory and in space [13, 14]. The problems range from recoverable anomalies, where the spacecraft is just out of service for a short period, to real failures rendering the spacecraft or part of it useless. Studies have indicated that charging is responsible for the majority of environment-related satellite anomalies, and of the charging-related anomalies the largest part is attributed to internal charging [15].

The most famous example concerns Telsat Canada's Anik E-1 and Anik E-2 communation satellites [16 – 18]. On 20 January 1994, after having been in orbit for about three years, first Anik E-1 began without warning to spin out of control (as shown by the headlines in Fig. 7.4). A few hours later the same anomaly happened to the second satellite Anik E-2. It was soon established that the gyroscopic guidance system, and specifically the momentum wheel control system, on both

**FIG. 7.4   Newspaper headlines after the Anik anomalies (image courtesy of L. J. Lanzerotti, Bell Laboratories, Lucent Technologies, Inc. © 1994.)**

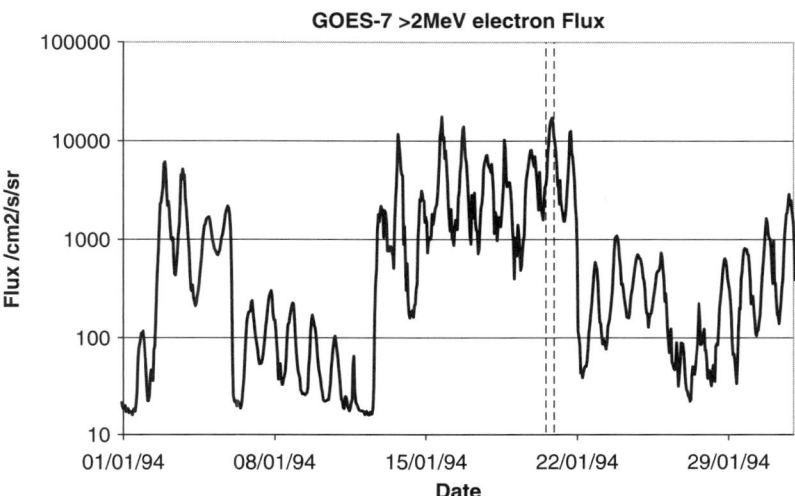

**FIG. 7.5    GOES-7 > 2 MeV electron fluxes in January 1994. Times of the Anik anomalies are shown by ——- (GOES data courtesy of NOAA/SEC).**

satellites had failed. This caused interruption of cable TV, telephone, and data-transfer services throughout Canada. With the help of a backup guidance system, the operators succeeded in bringing Anik E-1 back into service after about eight hours; however, the backup system on Anik E-2 failed. At first the total loss of Anik E-2 was feared. It was not until seven months later that Telsat managed to bring the satellite back into service. They had by then developed some workaround in the form of an innovative thrusters-based control system to maintain the correct pointing of the spacecraft. The anomalies themselves happened after a prolonged enhancement (more than one week long) of energetic (>2 MeV) electron fluxes at the geostationary altitude, where the satellites were located, as shown by data from the geostationary GOES-7 satellite in Fig. 7.5. After a thorough investigation internal charging was determined to be the only likely cause.

Another case that has been analyzed thoroughly and is taken as one of the first conclusive evidences of anomalies being caused by internal charging concerns the British DRA-δ spacecraft [13]. DRA-δ, in its geostationary orbit, experienced one to two anomalies per month over a period of several years in the early 1990s. None of the anomalies was a threat to the mission, but annoying mode switches changed the state of some telemetry channels. The clear conclusion from the subsequent investigation was that it was a question of internal dielectric charging. The anomalies did not occur preferentially at midnight to dawn local times, as would be expected of surface charging, but showed a clear correlation with longer periods of enhanced flux of high-energy electrons. Specifically an anomaly occurred almost every time that the daily fluence of $>2$ MeV electrons exceeded a threshold of $5 \times 10^7 \, \mathrm{cm}^{-2} \, \mathrm{day}^{-1} \, \mathrm{sr}^{-1}$.

Another case is the German scientific satellite Equator-S (data available online at http://www.mpe.mpg.de/EQS/), which was in a highly elliptic Earth orbit (570 $\times$ 65,693 km, inclination 7 deg). The satellite failed definitively in May 1998, after half a year in orbit instead of the planned two-year mission. Four months earlier one of two identical master control units had failed, and in May the second one also failed. Thanks to the redundant system good information on what went wrong could be obtained from the first failure. The conclusion of the following investigation was that the failure was attributed to a latch-up in one of a couple of specific components in the units possibly caused by internal charging. The times of both the failures were strongly correlated with long periods of enhanced flux of high-energy electrons.

Two other examples have been more controversial. The causes of the failures of the commercial geostationary satellites Telstar 401 in January 1997 and Galaxy 4 in May 1998 are still disputed [19]. For the Telstar 401 the operators suddenly lost contact with the spacecraft (interrupting the broadcast of a number of major U.S. television networks) and have not been able to communicate with it since. For Galaxy 4, which was a heavily used communications satellite (including a pager service with 45 million customers), the attitude control system suddenly failed. A backup system also failed at the same time (or maybe it had failed earlier), and so it became impossible to maintain a stable communication link. For Galaxy 4 the fluences of $>2$ MeV electrons were two to three orders of magnitude higher than the "normal" level for a period of at least two weeks just prior to the failure, which strongly suggests internal charging as the cause. Many experts in the space charging field regard the most likely cause of both of these failures to be internal charging, although the owners and operators of the spacecraft at the time believed this not to be the case.

The CRRES satellite, which carried an internal charging experiment [20], itself experienced some temporary anomalies coincident with the measurement of transients in the samples of the experiment [21]. The occurrence of the anomalies correlated well with high levels of high-energy electron flux and poorly with other hazard-related environmental parameters.

One of the most demanding environments with respect to internal charging is the environment of Jupiter, with its extremely intense radiation belts. These are dominated by electrons, and as well as having fluxes that are much higher than at the Earth they also include higher energies. It is believed that Jupiter's severe environment caused at least 42 internal charging discharge events during the single flyby by the Voyager 1 spacecraft of the planet in 1977 [22]. These events were power-on resets in the flight data subsystem and are believed to be caused by internal charging of a cable that built up enough charge to cause arcing. Consequently, internal charging was a major concern for the follow-on mission Galileo, which arrived in 1995 and spent about two years in the vicinity of the planet [23]. Most of this time was spent at a safe distance outside the radiation belts, but it had about 30 flybys of Jupiter and the various moons, including Europa, which is located at the heart of the radiation belts of Jupiter. The duration of each pass through the belts was of the order of two days. Because no obvious internal charging events were observed during this time, it appears that mitigation efforts were successful. In the design, stringent guidelines had been followed. These limited the use of highly resistive dielectrics, the size of isolated conductors, and imposed thick spacecraft shielding. A couple of years later these were formalized in the NASA handbook on avoiding problems caused by spacecraft on-orbit internal charging effects [24]. Currently, even more ambitious missions to Jupiter are being planned by NASA and ESA, putting an even higher demand on the mitigation efforts related to internal charging.

Details of the causes of anomalies and failures are often difficult to obtain from satellite owners. This is not simply a question of commercial confidentiality. Even when an anomaly has been tracked down to a specific unit or part of a unit, it is still very difficult to reproduce the exact same conditions in ground tests. This has led to some in-orbit and ground experiments where the conditions can be better controlled.

## 7.5  EXPERIMENTS IN SPACE AND ON GROUND

The CRRES spacecraft carried a dedicated instrument to measure ESDs arising from internal charging in space. The internal discharge monitor (IDM) [20] carried 16 insulator samples, including circuit boards and insulated wires. These were shielded from ambient plasma by 0.2 mm aluminum, and so any discharges seen on these samples must have been caused by internal charging. During the 14-month flight, IDM detected 4300 discharges, demonstrating the high sensitivity to internal charging of some dielectrics in the absence of protection measures. The pulse rate approximately followed the orbit-averaged electron flux, although pulses could occur hours after passing through the most intense part of the belt. From this experiment, a critical flux level of $5.5 \times 10^5 \ cm^{-2}s^{-1}$ was identified above which the risk of pulsing was high [25].

The ability of an internal charge ESD to trigger an anomaly was investigated in a ground-based study [6, 26] in which a printed circuit board was irradiated. It carried a dual D-type latch (also known as a flip-flop), which is a common building block in many digital circuits. Simultaneous with the detection of ESDs, many spontaneous changes of state of these latches were detected as shown in Fig. 7.6. The occurrence of these flips was shown to be associated with high charging current and low temperature.

## 7.6  SPACE WEATHER INFLUENCES

Electron fluxes in the outer radiation belt are highly dynamic, frequently varying by two or three orders of magnitude in a period of a few hours. As a result, internal charging currents are also highly variable. The variability of the outer belt is illustrated by Fig. 7.7, which shows enhancements seen by the SREM instrument on Proba-1 in a LEO polar orbit.

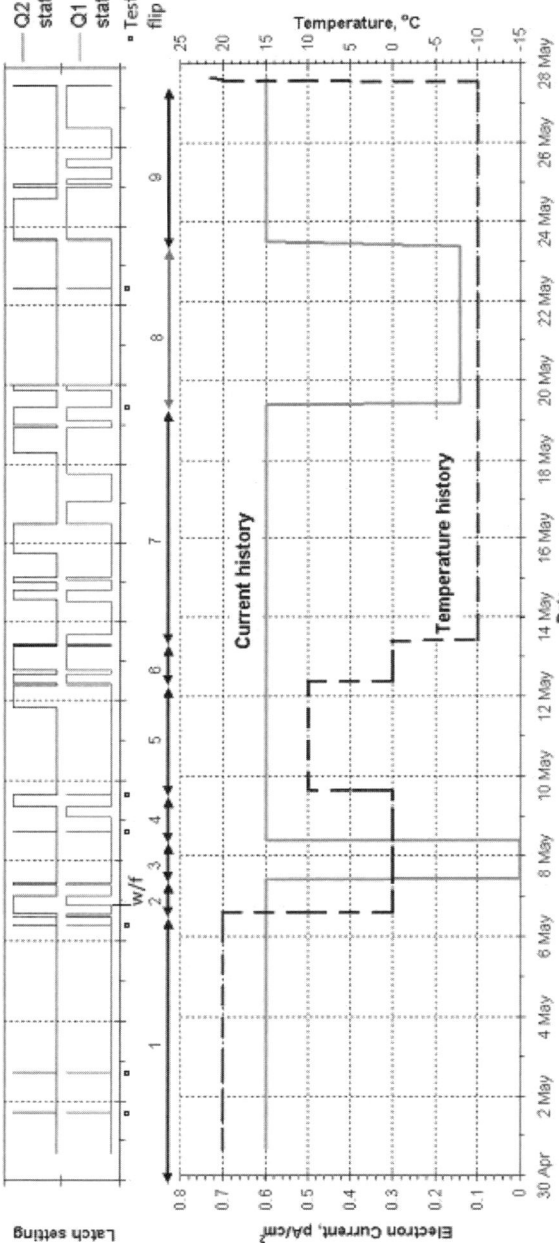

FIG. 7.6   The top panel shows the two output states from the two latches (Q1 and Q2) during testing. Some "test flips" are marked when the correct operation of the circuits was tested. Other spontaneous changes in state can be seen resulting from ESDs. The lower panel describes the irradiation current and temperature during the tests [6].

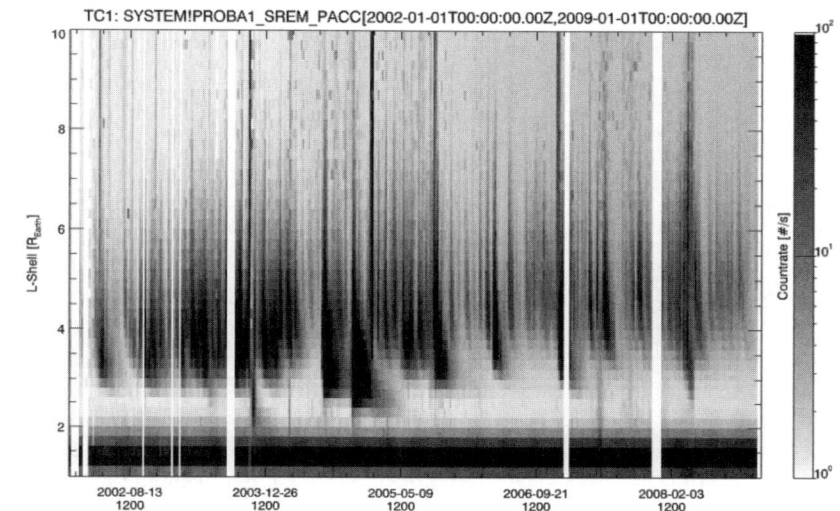

**FIG. 7.7    Counts in the Proba-1 SREM TC1 channel (>2 MeV electrons, >27 MeV protons) for 2002 to 2009 (inclusive), shown as a function of L-shell and time. Variations in count rate above L = 2.5 are due to changes in the intensity of the electrons in the outer radiation belt. The greyscale has been truncated at $10^2$.**

Similar variability is seen in the results of the Merlin radiation detector on MEO Giove-A [27]. This instrument is capable of making direct measurements of internal charging currents. Giove-A is in a 56-deg inclination, 23,200-km altitude circular orbit. This takes the spacecraft back and forth through the heart of the outer radiation belt. A summary plot of charging currents in three aluminum collector plates, each under a different thickness of aluminum shielding, is shown in Fig. 7.8. This exhibits a monthly repetition due to regular features on the sun that return with each solar rotation.

Enhancements in the outer radiation belt are usually associated with high-speed streams in the solar wind, which are emitted at coronal holes. The occurrence of these streams is most common in the declining phase of the solar cycle, and so the outer belt and the internal charging currents are most intense in this period.

In designing spacecraft equipment, the worst-case environment that is likely to be experienced must be taken into consideration. This requires a radiation belt model that is very different from long-term average models like AE8, which are used in dose analysis. Ideally, we want a model that shows the highest fluxes expected in the time defined by the material charging constant, that is, a different model for different materials.

In the FLUMIC model [28], a worst-case one-day average spectrum is proposed for all locations in the magnetosphere. This one-day timescale is used

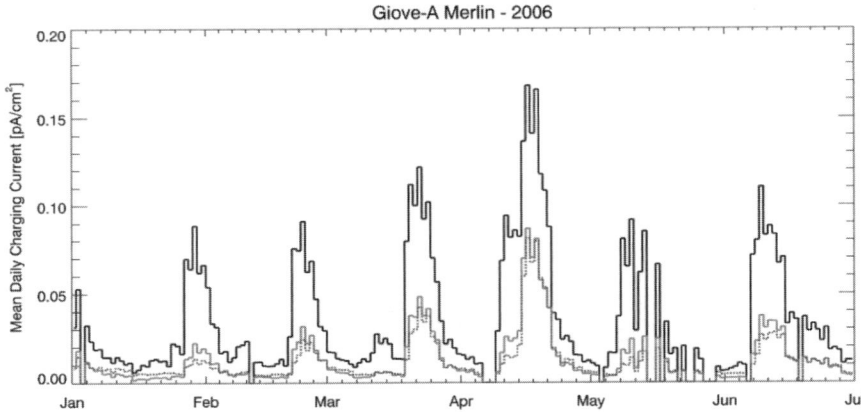

**FIG. 7.8    Internal charging current density recorded in MEO from January to July 2006 by the three plates in the Giove-A/Merlin detector. Daily average current densities are plotted as follows: black line—0.5-mm shield, 0.5-mm collector; grey line—1.0-mm shield, 0.5-mm collector; and grey dotted line—1.5-mm shield, 0.5-mm collector.**

because materials with shorter charging time constants are usually sufficiently conductive to avoid serious internal charging issues, and the one-day average is worst case for longer time constants. This model gives worst-case spectra, as a function of time of year and time during the solar cycle, throughout the inner and outer belts. This model has been adopted by ESA and is part of the ECSS Space Environment standard [29]. NASA has issued a worst-case environment for geostationary orbit [24]. This is the spectrum corresponding with a 99.9 percentile of >2 MeV fluxes, as seen by GOES spacecraft. Similarly, Wrenn et al. [30] and Fennel et al. [31] have published worst-case spectra for geostationary orbit. Figure 7.9 compares these models. (Note that the maximum and minimum values for FLUMIC correspond nominally to 14 April 1994 and 21 December 1989 respectively for a geostationary orbit.)

The resulting charging currents from these models are shown in Fig. 7.10. Most spacecraft skins provide at least 1-mm Al equivalent shielding, and so charging currents are usually around 0.1 pA/cm$^2$ or lower.

Although the Jovian radiation environment has been less extensively measured than that of the Earth, environment models have been developed over a number of years [32–34]. The electron spectrum expected near the moons Ganymede and Europa is shown in Fig. 7.11 [32].

## 7.7   MATERIAL DEPENDENCE

An electron's penetration depth is dependent on its energy and the shielding properties of the material through which it passes, particularly its density.

Electrons interact mostly with the bound electrons in atoms, and the number of electrons per atoms is approximately a constant factor of atomic number, except for heavy elements (above Fe). Hence, the approximation can be made that shielding efficiency is proportional to mass density, except for high-Z materials, such as tungsten shielding. The range formula of Weber [1] shown earlier is therefore a reasonable approximation for low-to-medium atomic number materials, if scaled by density.

The differences in materials are seen most strongly in their conductivity. Metals and semiconductors are able to conduct charge away before hazardous potentials can arise. Even among insulators, there are strong variations in intrinsic conductivity and the dependence of conductivity on environmental factors.

Temperature dependence of conductivity in insulators is usually expressed by the Arrhenius equation [35]:

$$\sigma(T) = \sigma_\infty \exp\left(-\frac{E_A}{kT}\right)$$

where the activation energy $E_A$ and $\sigma_\infty$ are constants. Almost all electrons belonging to molecules in insulators are trapped in potential wells with energy well below the conduction band. Only in the very high-energy tail of the distribution are a few

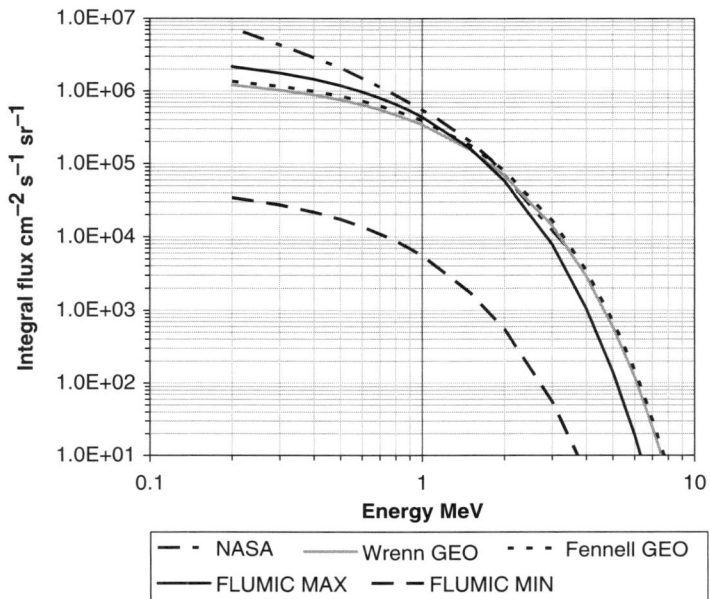

FIG. 7.9  Comparison of the FLUMIC [28] model with the NASA [24], Wrenn GEO [30], and Fennell GEO [31] worst-case spectra.

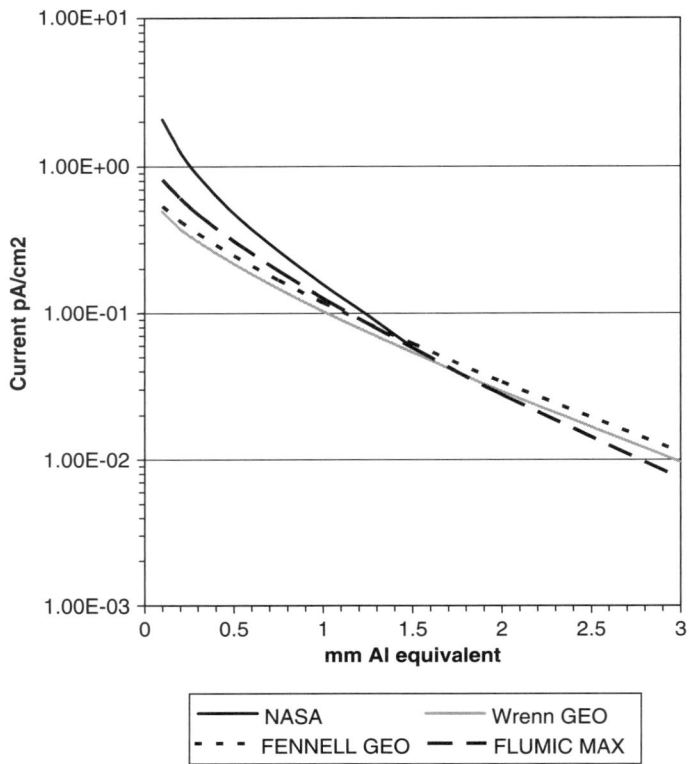

**FIG. 7.10  Comparison of charging currents as a function of shielding thickness for the NASA [24], FLUMIC MAX [28], Wrenn [30], and Fennel [31] geostationary worst-case spectra.**

electrons able to escape and migrate under the influence of external electric fields. However, as temperature rises, this tail increases exponentially, and so conductivity similarly increases. As a result, heating dielectrics is a viable way to control internal charging [36] although it has not yet been widely exploited.

Sim and Dennison [48] have recently described a combined theory of electric field and thermal-induced changes in conductivity, based also on concepts of charge carried hopping between quantum states. This indicates that the conductivity should be proportional to exp $(T^{-1})$ or exp $(T^{-1/4})$ depending on temperature range.

In high electric fields, insulators start to conduct more strongly. This effect was described quantitatively by Adamec and Calderwood [37].

$$\sigma(E,\,T) = \sigma(T)\left[\frac{2 + \cosh\left(\beta_F E^{1/2}/2kT\right)}{3}\right]\left[\frac{2kT}{eE\delta}\sinh\left(\frac{eE\delta}{2kT}\right)\right]$$

where $\beta_F = \sqrt{e^3/\pi\varepsilon}$, $\delta$ is a distance that expresses how far charge carriers can jump over a potential barrier, $e$ is the charge on an electron, and $\epsilon$ is permittivity. This equation was theoretically derived in [38] to describe the way the electric field increases the mobility of carriers and causes the activation of additional carriers. However a fixed value of $\delta(10$ Å) was chosen to fit with experimental data.

Conductivity is also enhanced by the extra carriers created due to ionizing radiation. This dose-rate enhanced conductivity is described by the Fowler formula [39]:

$$\sigma = \sigma_0 + k_p \dot{D}^\Delta \; \Omega^{-1} \; cm^{-1}$$

where $\sigma_0$ is the conductivity in the absence of irradiation (called "dark conductivity"), $k_p$ is a material-dependent coefficient of prompt radiation-induced conductivity, and $\Delta$ is a dimensionless material-dependent exponent, usually between 0.6 and 1.0 for most polymers.

It is generally observed that after irradiation is stopped radiation-induced conductivity decays away only slowly [6]. This called "delayed" radiation-induced conductivity, and in some cases a dose-dependent permanent increase in conductivity is also thought to occur. However, this phenomenon seems to be less common and might be related to permanent changes under irradiation for some particular materials.

Measurement of conductivity in dielectric materials is generally carried out according to international standards such IEC-93 [40] or ATSM-D257 [41]. In

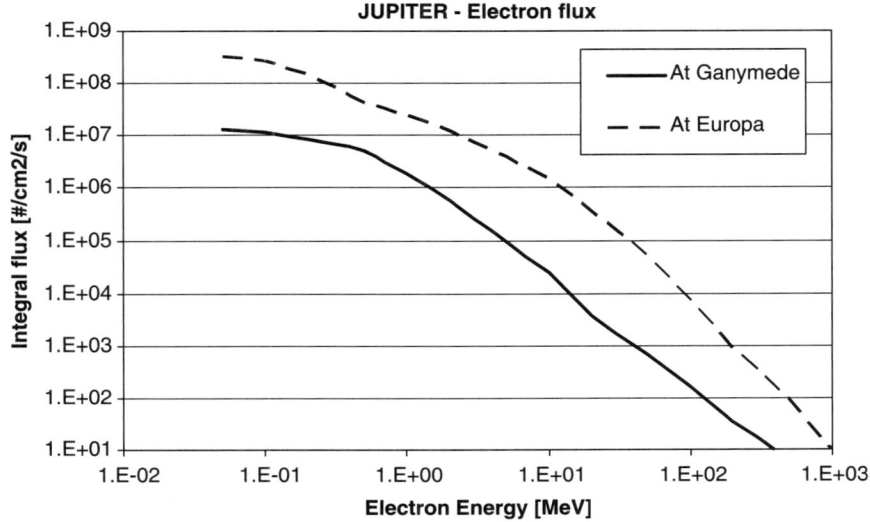

**FIG. 7.11   Mean electron flux spectra in the Jovian environment [32].**

these tests the material is sandwiched between parallel plates, and a measurement is made of current passing through the material. Usually 60 s are allowed for the current to stabilize. However, for good insulators the intrinsic conductivity can be exceeded by temporary polarization currents in this timescale, and so the tests exaggerate the true conductivity. Hence, it is necessary to carry out this measurement over a timescale representative of the internal charging process, for example, $>1$ day.

In an alternative method of determining conductivity, a low-energy electron beam ($\sim 10$ keV) is used to charge a sample of the material, and the exponential decay of the resultant surface potential is used to infer the material time constant, that is, $V = V_0 \exp\left(-t/\tau\right)$. Hence, the conductivity can be deduced if the dielectric constant is known.

A method of simultaneously deriving up to five of the dielectric material properties ($K_p$, $\Delta$, $\epsilon$, $E_A$, and $\sigma_0$) has been developed [41]. This incorporates the DICTAT simulation tool [42] that models internal charging. A series of high-energy electron irradiation tests is performed covering different intensities, beam energies, and temperatures, while the surface potential is repeatedly measured. The method has been coded into a tool called DICFIT, which works in the following way.

DICTAT can calculate the surface potential, that is, the same parameter that is measured in the tests. It can be considered to be a nonlinear mathematical function [DICTAT($u$, $x$)], where $u$ describes the material and $x$ is a state vector describing the test. Here $x$ is defined to be beam intensity, shield thickness, dielectric thickness, temperature, and irradiation time. If all five material parameters are unknown, then $u$ consists of $K_p$, $\Delta$, $\epsilon$, $E_A$, and $\sigma_0$. Also, $u$ can contain a subset of these elements if some parameters are known a priori. Each laboratory voltage measurement $V$ and corresponding DICTAT simulation provide us with an equation of the form:

$$\frac{\text{DICTAT}(u, x)}{V(x)} - 1 = fn(u, x) = 0$$

If we have five equations like this, that is, five experimental measurements, we can solve exactly for the five material parameters as long as the measurements and equations are perfectly accurate. However, because of unavoidable errors in the measurements and uncertainties in the theoretical equations, it is better to have many more measurements, covering a wide range of environment parameters $x$. These are solved using the multidimensional Newton–Raphson method. This follows the local gradient ($dfn/du$) starting from an initial guess and converges to a minimum. At each iteration, it calculates the local gradient using a multidimensional least-squares fit. This is performed using the singular value decomposition method, which is able to handle cases where one or more parameters make no significant contribution to the result (i.e., degeneracy).

As a result, DICFIT determines which material parameters best explain the observations. It was employed to generate a database of internal charging-related material parameters for some common space materials [9] listed in Table 7.1, based on irradiation tests representative of the GEO environment.

## 7.8  GEOMETRIC EFFECTS

Some dielectric structures can be considered to be essentially planar, for example, printed circuit boards and multilayer insulation. Assuming uniform irradiation, it is easiest to calculate the electric field structure in an infinite plane for a structure like that shown in Fig. 7.12.

The electric field for the equilibrium case (also the worst case) is given by

$$E = j/\sigma$$

where $E$ is the maximum field and $j$ is the local conducted current. The voltage is given by the integral of the electric field over distance $d$ normal to the plane's surface,

$$V = \int_0^d E.\mathrm{d}x = \int_0^d j(x)/\sigma(x).\mathrm{d}x = \frac{I}{A}\frac{RA}{d}\mathrm{d} = IR$$

because $I/A$ is the integral of the locally deposited current density $j$ and $RA/d$ is the integral of $1/\sigma$. For a planar structure with one grounded surface, the maximum $j$ and therefore $E$ are seen next to this surface. For a planar structure with grounding on both surfaces, the value of $j$ can be reduced by as much as $\frac{1}{2}$ because current can flow to both surfaces.

For complicated three-dimensional structures, the current can concentrate or diverge as a result of the shape of the structure, but the preceding equation in terms of the electric field will hold locally, and the voltage is the three-dimensional integral along the electric field vector.

An example of a three-dimensional structure is a connector with a large dielectric volume containing small metallic pins. Because all of the current collected in the dielectric will eventually flow to the pins with small surface area, the internal field can be greatly enhanced compared to a planar structure of similar thickness. This is illustrated in two dimensions in Fig. 7.13.

## 7.9  SIMULATION TOOLS

For engineering design, knowledge of the maximum internal electric field and maximum surface voltage, expected on a structure during the mission, is important. These have to be kept below thresholds for electrostatic breakdown (internal

**TABLE 7.1   MATERIAL PARAMETERS DETERMINED FROM FITTING OF EXPERIMENTAL RESULTS [9]**

| Material | Density, $g/cm^3$ | Dark Conductivity, $(\Omega^{-1}m^{-1})$ | Dielectric Constant | $K_p$, $(\Omega^{-1}m^{-1}rads^{-\Delta}s^{\Delta})$ | $\Delta$ | $E_A$, eV | Charging Hazard[a] |
|---|---|---|---|---|---|---|---|
| Betacloth | 1.05 | $1.46 \times 10^{-15}$ | 3.2 | n/s[b] | n/s | 2.50 | Low |
| CFRP[c] | 1.1 | $>3.11 \times 10^{-13}$ | – | – | – | – | V low |
| Delrin | 1.41 | $4.41 \times 10^{-14}$ | (4.0) | n/s | n/s | 1.26 | Low |
| FEP | 2.15 | $2.78 \times 10^{-16}$ | 2.91 | $3.91 \times 10^{-15}$ | 0.36 | 0.25 | High |
| FR-4 | 2.06 | $8.48 \times 10^{-16}$ | 5.59 | $1.73 \times 10^{-20}$ | 1.07 | 2.44 | High |
| LDPE | 0.92 | $6.94 \times 10^{-15}$ | 4.26 | $6.97 \times 10^{-14}$ | 1.08 | 1.16 | Low |
| PMMA | 1.19 | $3.05 \times 10^{-17}$ | 3.95 | n/s | n/s | 0.47 | High |
| Polyimide | 1.42 | $1.49 \times 10^{-16}$ | 3.01 | n/s | n/s | 1.75 | High |
| POM | 1.41 | $1.54 \times 10^{-14}$ | 3.72 | $2.07 \times 10^{-13}$ | 1.57 | 1.11 | Low |
| Solithane | 0.91 | $3.56 \times 10^{-14}$ | 12.47 | $1.73 \times 10^{-15}$ | 0.57 | 1.33 | Low |

a   Charging hazard is an assessment of whether these material parameters, combined with a worst-case GEO environment and minimal shielding, will generally lead to electric fields above 10 MV/m. High-risk materials are characterized by low conductivity, generally less than $10^{-15}$ $\Omega^{-1}m^{-1}$.

b   n/s indicates that a parameter is not significant and can be treated as zero.

c   CFRP was so conductive that only a lower limit for the conductivity could be deduced.

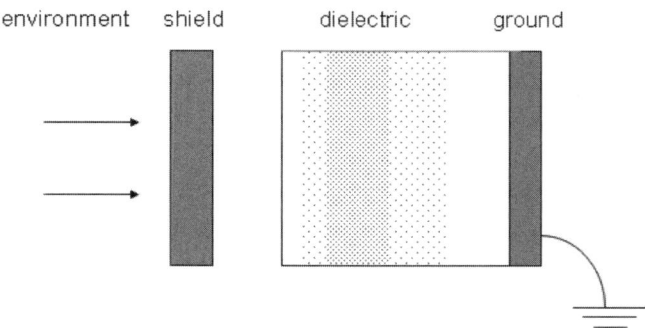

**FIG. 7.12   Schematic section of a planar dielectric (infinite in dimensions perpendicular to the irradiation). Dots indicate the nonuniform deposition of electrons from the environment.**

punch-through) and surface discharge (blowoff); otherwise, there is a danger of upsets and damage to electronic components.

There are a few one-dimensional codes that can be used for assessing internal charging levels:

- ESADDC [44] uses the one-dimensional Monte Carlo radiation transport code from the ITS radiation suite [2] to calculate charge deposition in a set of planar layers that can be of the same or differing materials. A circuit solver handles the conductivity and capacitance between adjacent layers. The effects of electric-field-induced conductivity and radiation-induced conductivity are included.

- DICTAT [43] was developed as a simpler and quicker code, with the aim of performing the same types of calculations for engineering applications. (ESADDC, although more scientifically accurate, is slower, harder to set up

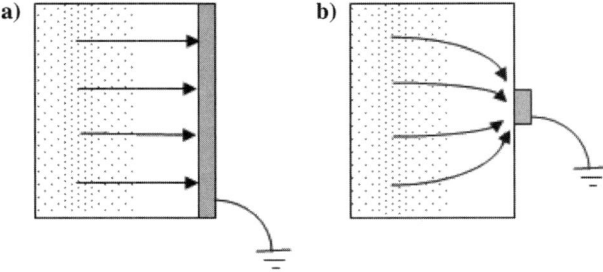

**FIG. 7.13   Illustration of the concentration of conducted current (shown by black arrows): a) the simple planar case and b) the case of an area around a small grounded pin. It can be seen that near the pin the current density $j$ increases and therefore so does electric field $E$.**

and not easily portable.) DICTAT uses the Weber [1] and Sorensen [5] equations to perform a simple analytical calculation for electron transport through aluminum, and it takes advantage of the fact that for most materials the shielding is the same as for aluminum as long as it is scaled by the material density. The geometries simulated are one-dimensional planar or cylindrical. DICTAT also takes into account modification of conductivity by temperature, radiation dose rate, and electric field. Several sets of material parameters are included as defaults. DICTAT is accessible as part of the SPENVIS [45] web service. A number of different environment inputs are possible: a user-defined static spectrum and a time-varying sequence of spectra or a spectrum automatically calculated from the FLUMIC model. DICTAT is capable of performing iterative calculations to calculate changes to shielding or dielectric thickness needed to avoid dangerous charging levels.

- NUMIT [46] is a one-dimensional bulk charging tool that calculates the fields in a dielectric held between two metallic plates. The code is capable of being easily modified, so that other geometries can be simulated. It can accept static spectra and time-dependent environments. Like DICTAT, it uses an analytical electron transport formulation (the method developed by Frederickson and Bell [7]) to give the deposited charge and takes into account the capacitance and conductivity, including radiation-induced conductivity.

There are various codes that are capable of determining the three-dimensional charge deposition in dielectrics. These are principally three-dimensional Monte Carlo radiation transport codes, such as Geant-4 [47]. Up to now, no one has combined such a code with both a three-dimensional calculation of electric fields and of the current flow due to conductivity. The development of such a tool is planned as part of a new ESA study into electron radiation effects.

## 7.10  MITIGATION

It is part of the design process to prevent hazardous internal charging levels being reached, and there are a number of steps that can be taken. Metallic components can be excluded as a source of ESD simply by grounding them. Grounding via a bleed resistor is generally sufficient. For dielectric materials, there is no simple solution, but there are a number of approaches that can be taken:

- Control of environment
  - Avoiding hazardous regions – Choosing a more benign orbit is sometimes possible. This might also reduce radiation doses and so help reduce dose-related effects.
- Control of deposited current
  - Thick shields – This reduces the total current reaching the material.
  - Thin dielectrics – This makes the deposited current less because the difference between the entering and exiting currents is smaller. There is the

added advantage that surface voltage, which is the integral of the electric field, is also reduced. Adding conductive layers, for example, in multilayer printed circuit boards, does effectively the same thing by breaking a thick dielectric into several thin layers.

- Control of dielectric conductivity
  - o Leaky insulators – Even a small leakage current can effectively remove the internal charging threat. Carbon-doping is sometimes employed to improve the conductivity of dielectrics.
  - o High temperatures – Because the conductivity of dielectrics is usually very sensitive to temperature, raising the temperature can be very effective. If charging is expected over a long timescale, occasional increases in temperature can be used to periodically discharge the dielectric.
- Control of circuit sensitivity
  - o Slow electronic circuits – Because ESDs cause transients of typically tens of nanoseconds, using slower components would make a circuit less sensitive. However, the opposite trend is happening, with faster, more sensitive devices becoming more common.
  - o Pulse filtering and damping resistors - The sensitivity of circuits to ESD can be modified by introducing filters that diminish the effects of the resulting short timescale transients. This cannot be applied everywhere but could, for example, be used on a power line.
- Control of software sensitivity
  - o Error detection and correction – If upsets to electronics cannot be avoided, it is possible to make data systems more resilient by correcting random errors.

Guidelines for assessing the internal charging risk and for designing a spacecraft free from internal charging problems have been published in NASA Handbook NASA-HBDK-4002 "Avoiding Problems Caused by Spacecraft On-Orbit Internal Charging Effects" [24]. Under the European Cooperation for Space Standardization (ECSS) framework, a charging standard (ECSS-E-ST-20-06C Spacecraft Charging) [47] that includes requirements for the assessment and limitation of internal charging effects has been issued.

## 7.11  CONCLUSION

Internal charging occurs when high-energy electrons are deposited in dielectrics and ungrounded metallic components. This can occur in surface layers such as exposed cables and multilayer insulation, but the main concern is usually with equipment inside the spacecraft Faraday cage where ESD has easier access to sensitive components. This process has been the cause of many anomalies. Trapped electrons in the radiation belt are the principal cause near the Earth, and similar populations in the magnetospheres of Jupiter and Saturn also represent a threat.

Charging levels are dependent on the environment, shielding, and the shape and electrical properties of the material in question. Analysis tools are available to assess this problem in one dimension, but currently there is a lack of three-dimensional models that can tackle all aspects of the process.

As our understanding of this process has grown, improved mitigation techniques, encapsulated in a NASA Handbook and an ECSS Standard, are leading to better mitigation. However, advances in technology (faster and smaller devices) are leading to increased sensitivity, and internal charging will remain a significant hazard in the foreseeable future.

# REFERENCES

[1]   Weber, K. H., "Eine Einfache Reicheweite-Energie-Beziehung fuer Electronen im Energiebereich von 3 keV bis 3 MeV," *Nuclear Instruments and Methods*, Vol. 25, 1964, p. 261.

[2]   Halbleib, J. A., Kensek, R. P., Mehlhorn, T. A., Valdez, G. D., Seltzer, S. M., and Berger, M. J., "ITS version 3.0: The Integrated TIGER Series of Coupled Electron/ Photon Monte Carlo Transport Codes," Sandia National Lab., Albuquerque, NM, March 1992.

[3]   GEANT Detector Description and Simulation Tool, CERN Program Library Long Writeup W5013, CERN, Geneva, Switzerland, 1993.

[4]   Trenkel, C., "Comparison of GEANT 3.15 and ITS 3.0 Radiation Transport Codes," ESA Working Paper, EWP 1747, Oct. 1993.

[5]   Sørensen, J., "An Engineering Specification of Internal Charging," *Environment Modelling for Space-Based Applications*, edited by A. Hilgers and T. D. Guyenne, ESA SP-392, ESA, Noordwijk, The Netherlands, 1996, p. 129.

[6]   Rodgers, D. J., Ryden, K. A., Latham, P. M., Levy, L., and Panabiere, G., "Engineering Tools for Internal Charging," ESA, Final Rept. ESA Contract 12115/96/NL/JG(SC), DERA/CIS(CIS2)/7/36/2/4/FINAL, Noordwijk, The Netherlands, 1999.

[7]   Frederickson, A. R., and Bell, J. T., "Analytic Approximation foe Charge Current and Deposition by 0.1 to 100 MeV Electrons in Thick Slabs," *IEEE Transactions on Nuclear Science*, Vol. 42, No. 6, 1995, p. 1910.

[8]   Tabata, T., and Ito, R., "An Algorithm for the Energy Deposition by Fast Electrons," *Nuclear Science and Engineering*, Vol. 53, No. 2, 1974, pp. 226–239.

[9]   Rodgers, D. J., Ryden, K. A., Bielby, R., Hunter, K. A., Clucas, S. N., Morris, P. A., Wrenn, G. L., and Levy, L., "Internal Charging Anomaly Study and Tool Development," ESA, Final Rept. ESA Study 16265/02/NL/FM, QINETIQ/KI/ SPACE/TR031642, Noordwijk, The Netherlands, Feb. 2005.

[10]  Knudson, A. R., Buchner, S., McDonald, P., Stapor, W. J., Lewis, S., and Zhao, Y., "The Effects of Radiation on MEMS Accelerometers," *IEEE Transactions on Nuclear Science*, Vol. 43, No. 6, 1996, pp. 3122–3126.

[11]  Touboul, P., Foulon, B., and Willemenot, E., "Electrostatic Space Accelerometers for Present and Future Missions," *Acta Astronautica*, Vol. 45, No. 10, 1999, pp. 605–617.

[12]   Araujo, H. M., Wass, P., Shaul, D., Rochester, G., and Sumner, T. J., "Detailed Calculation of Test-Mass Charging in the LISA Mission," *Astroparticle Physics*, Vol. 22, Nos. 5–6, 2005, pp. 451–469.

[13]   Wrenn, G. L., "Conclusive Evidence for Internal Dielectric Charging Anomalies on Geosynchronous Communications Spacecraft," *Journal of Spacecraft and Rockets*, Vol. 32, No. 3, 1995, pp. 514–520.

[14]   Fredrickson, A. R., "Upsets Related to Spacecraft Charging," *IEEE Transactions on Nuclear Science*, Vol. 43, No. 2, April 1996, pp. 426–441.

[15]   Fennell, J. F., Koons, H. C., Roeder, J. L., and Blake, J. B., "Spacecraft Charging: Observations and Relationship to Satellite Anomalies," Aerospace Corp., Rept. TR-2001(8570)-5, El Segundo, CA, Aug. 2001.

[16]   Knapp, B., "Telsat Ponders Using Thrusters to Salvage ANIK," *Space News*, Vol. 5, No. 5, 1994, p. 1.

[17]   Hughes, D., "Telsat Succeeds in Anik E2 Rescue," *Aviation Week and Space Technology*, Vol. 141, No. 1, 4 July 1994, p. 32.

[18]   Leach, R. D., and Alexander, M. B., "Failures and Anomalies Attributed to Spacecraft Charging," NASA Reference Publication 1375, Aug. 1995.

[19]   Baker, D. N., "The Occurrence of Operational Anomalies in Spacecraft and Their Relationship to Space Weather," *IEEE Transactions on Plasma Science*, Vol. 28, Dec. 2000, pp. 2007–2016.

[20]   Frederickson, A. R., Mullen, E. G., Kerns, K. J., Robinson, P. A., and Holeman, E. G., "The CRRES IDM Spacecraft Experiment for Insulator Discharge Pulses," *IEEE Transactions on Nuclear Science*, Vol. 40, No. 2, 1993, p. 233.

[21]   Violet, M. D., and Frederickson, A. R., "Spacecraft Anomalies on the CRRES Satellite Correlated with the Environment and Insulator Samples," *IEEE Transactions on Nuclear Science*, Vol. 40, No. 6, Dec. 1993, pp. 1512–1520.

[22]   Evans, R. W., and Garrett, H. B., "Modelling Jupiter's Internal Electrostatic Discharge Environment," *Journal of Spacecraft and Rockets*, Vol. 39, No. 6, 2002, pp. 926–932.

[23]   Fieseler, P. D., Ardalan, S. M., and Frederickson, A. R., "The Radiation Effects on Galileo Spacecraft Systems at Jupiter," *IEEE Transactions on Nuclear Science*, Vol. 49, No. 6, Dec. 2002, pp. 2739–2758.

[24]   Whittlesey, A., "Avoiding Problems Caused by Spacecraft On-Orbit Internal Charging Effects," NASA Technical Handbook, NASA-HDBK-4002, 17 Feb. 1999.

[25]   Frederickson, A. R., Holeman, E. G., and Mullen, E. G., "Characteristics of Spontaneous Electrical Discharging of Various Insulators in Space Irradiations," *IEEE Transactions on Nuclear Science*, Vol. 39, No. 6, 1992, pp. 1773–1782.

[26]   Ryden, K. A., Morris, P. A., Rodgers, D. J., Bielby, R., Knight, P. R., and Sorensen, J., "Improved Demonstration of Internal Charging Hazards Using the 'Realistic Electron Environment Facility' (REEF)," *Proceedings of the 8th Spacecraft Charging Technology Conference*, edited by J. L. Minor, NASA/CP-2004-231091, NASA, Marshall Space Flight Center, Huntsville, AL, 2003.

[27]   Ryden, K. A., Morris, P. A., Ford, K. A., Dyer, C. S., Taylor, B., Underwood, C. I., Jason, S., Rodgers, D., Mandorlo, G., Gatti, G., Evans, H. D., and Daly, E. J., "Observations of Internal Charging Currents in Medium Earth Orbit," 10th Spacecraft Charging Technology Conference, June 2007.

[28]  Rodgers, D. J., Hunter, K. A., and Wrenn, G. L., "The FLUMIC Electron Environment Model," *Proceedings of the 8th Spacecraft Charging Technology Conference*, edited by J. L. Minor, NASA/CP-2004-231091, NASA, Marshall Space Flight Center, Huntsville, AL, 2003.

[29]  "Space Environment," ECSS-E-ST-10-04C, Noordwijk, The Netherlands, 15 Nov. 2008, http://www.ecss.nl.

[30]  Wrenn, G. L., Rodgers, D. J., and Buehler, P., "Modelling the Outer Belt Enhancements of MeV Electrons," *Journal of Spacecraft and Rockets*, Vol. 37, No. 3, 2000, pp. 408–415.

[31]  Fennell, J. F., Koons, H. C., Chen, M. W., and Blake, J. B., "Internal Charging: A Preliminary Environmental Specification for Satellites," *IEEE Transactions on Plasma Science*, Vol. 28, No. 6, Dec. 2000, pp. 2029–2036.

[32]  Divine, N., and Garrett, H. B., "Charged Particle Distributions in Jupiter's Magnetosphere," *Journal of Geophysical Research*, Vol. 88, Sept. 1983, pp. 6889–6903.

[33]  Jun, I., Garrett, H. B., and Evans, R. W., "High-Energy Trapped Particle Environments at Jupiter: an Update," *IEEE Transactions on Nuclear Science*, Vol. 52, Dec. 2005, p. 2281.

[34]  Bourdarie, S., and Sicard-Piet, A., "Jupiter Environment Modelling," ONERA, Technical Note 120, Issue 1.2, ESA contract 19735/NL/HB, FR 1/11189 DESP, Noordwijk, The Netherlands, Oct. 2006.

[35]  Dissado, L. A., and Fothergill, J. C., *Electrical Degradation and Breakdown in Polymers*, IEE Materials and Devices Series 9, Peter Peregrinus, Ltd., London, 1992.

[36]  Ryden, K. A., Morris, P. A., and Clucas, S. N., "The Effect of Temperature on Internal Charging and Its Potential Role in Hazard Mitigation," *Proceedings of the 7th SCTC*, ESA-SP476, ESA, Noordwijk, The Netherlands, Nov. 2001, pp. 133–138.

[37]  Sim, A. M., and Dennison, J. R., "Parametrization of Temperature, Electric Field, Dose Rate and Time Dependence of Low Conductivity Spacecraft Materials Using a Unified Electron Transport Model," *Proceedings of the 11th Spacecraft Charging Technology Conference*, U.S. Air Force Research Lab., Albuquerque, NM, 2010.

[38]  Adamec, V., and Calderwood, J. H., "Electrical Conduction in Dielectrics at High Fields," *Journal of Applied Physics*, Vol. 8, No. 5, 1975, pp. 551–560.

[39]  Fowler, J. F., "X-Ray Induced Conductivity in Insulating Materials," *Proceedings of the Royal Society of London, Series A: Mathematical and Physical Sciences*, Vol. 236, No. 1207, 1956, pp. 464–480.

[40]  "Methods of Test for Volume Resistivity and Surface Resistivity of Solid Electrical Insulating Materials," IEC 60093, ed. 2.0, IEC, Geneva, Switzerland, 1980.

[41]  "Standard Test Methods for DC Resistance or Conductance of Insulating Materials," American Society for Testing and Materials, ASTM D 257-93, 1993 (Re-approved 1998).

[42]  Rodgers, D. J., Ryden, K. A., Wrenn, G. L., Lévy, L., and Sørensen, J., "Fitting of Material Parameters for DICTAT Internal Dielectric Charging Simulations Using DICFIT," *Proceedings of the 9th International Symposium on Materials in a Space Environment*, edited by K. Fletcher, ESA SP-476, ESA, Noordwijk, The Netherlands, 2003, p. 609.

[43]   Rodgers, D. J., Ryden, K. A., Wrenn, G. L., Latham, P. M., and Sørensen, J., "An Engineering Tool for the Prediction of Internal Dielectric Charging," *Proceedings of the 6th Spacecraft Charging Technology Conference*, AFRL-VS-TR-20001578, U.S. Air Force Research Lab., Hanscom AFB, MA, Sept. 2000, pp. 125–129.

[44]   Soubeyran, A., and Floberhagen, R., "ESADDC 1.1 User Manual," Matra Marconi Space, ESA Contract No. 9558/91/NL/JG-WO12, Toulouse, France, Feb. 1994.

[45]   Heynderickx, D., Quaghebeur, B., Speelman, E., and Daly, E., "Space Environment Information System (SPENVIS): A WWW Interface to Models of the Space Environment and Its Effects," AIAA Paper 2000-0371, Jan. 2000, www.spenvis.oma.be/ spenvis/.

[46]   Jun, I., Garrett, H. B., Kim, W., and Minow, J., "Review of an Internal Charging Code, NUMIT," *Proceedings of the 10th International Spacecraft Charging Conference*, ONERA, Toulouse, France, 2007.

[47]   Agostinelli, S., and GEANT Collaboration, "Geant4: a Simulation Toolkit," *Nuclear Instruments and Methods*, Vol. A506, No. 3, 2003, p. 250, http://cern.ch/geant4.

[48]   "Spacecraft Charging Standard," ECSS, ECSS-E-ST-20-06C, Noordwijk, The Netherlands, 31 July 2008, www.ecss.nl.

# INDEX

*Note*: Page numbers with "f" represent figures. Page numbers with "t" represent tables.

# SUPPORTING MATERIALS

Many of the topics introduced in this book are discussed in more detail in other AIAA publications. For a complete listing of titles in the AIAA Progress in Astronautics and Aeronautics series, as well as other AIAA publications, please visit www.aiaa.org. AIAA is committed to devoting resources to the education of both practicing and future aerospace professionals. In 1996, the AIAA Foundation was founded. Its programs enhance scientific literacy and advance the arts and sciences of aerospace. For more information, please visit www.aiaafoundation.org.